Resistance of Concrete to Chloride Ingress

T0227812

Chloride ingress in reinforced concrete is a major problem, inducing corrosion and consequent spalling and structural weakness. It occurs worldwide and imposes enormous costs. In order to guarantee the integrity of concrete structures and maintain them properly to achieve their designed service life, test methods and relevant models for service life prediction are required.

Setting out current understanding of chloride transport mechanisms, test methods and prediction models, this book:

- describes basic mechanisms and theories;
- classifies the commonly used parameters and units which express chloride and its transport properties;
- outlines laboratory and in-field test methods including precision results from inter-laboratory comparison tests;
- explains some of the fundamentals underlying the main models;
- carries out a theoretical appraisal of different models;
- uses analytical and probabilistic approaches for sensitivity analysis of various models;
- presents and discusses the results from a benchmarking evaluation of different models;
- gives guidelines for the practical use of test methods and models, including tests for *in situ* applications. Test methods validated by the precision results are detailed in an appendix.

Providing practising engineers, designers, researchers, advanced students and other professionals with a useful reference for the analysis and design of concrete structures exposed to chloride environments, the book draws on the CHLORTEST project, which involved seventeen partners from ten European countries, and will serve as an authoritative guide for some time to come.

Tang Luping is Professor at Chalmers University of Technology, Sweden.

Lars-Olof Nilsson is Professor at Lund Institute of Technology, Sweden.

P.A. Muhammed Basheer is Professor at Queen's University Belfast, UK.

Resistance of Concrete to Chloride Ingress
Testing and modelling

Tang Luping, Lars-Olof Nilsson
and P.A. Muhammed Basheer

CRC Press
Taylor & Francis Group
Boca Raton London New York

CRC Press is an imprint of the
Taylor & Francis Group, an **informa** business

A SPON PRESS BOOK

First published 2012
by Spon Press
2 Park Square, Milton Park, Abingdon, Oxfordshire OX14 4RN

Simultaneously published in the USA and Canada
by Spon Press
711 Third Avenue, New York, NY 10017

First issued in paperback 2017

Spon Press is an imprint of the Taylor & Francis Group, an informa business

British Library Cataloguing in Publication Data
A catalogue record for this book is available from the British Library

Library of Congress Cataloging in Publication Data
Tang, Luping, 1956–
    Resistance of concrete to chloride ingress : testing and modelling /
    Luping Tang, Lars-Olof Nilsson, and P.A. Muhammed Basheer.
    p. cm.
    Includes index.
    1. Reinforced concrete—Chemical resistance.
    2. Reinforced concrete—Testing. 3. Reinforced concrete—
    Corrosion—Simulation methods. 4. Chlorides.
    I. Nilsson, Lars-Olof. II. Basheer, P. A. Muhammad
    (Paliakarakadu Assen Muhammed) III. Title.
    TA445.T36 2011
    620.1'3723—dc23
    2011014014

ISBN 13: 978-1-138-07759-1 (pbk)
ISBN 13: 978-0-415-48614-9 (hbk)

Typeset in Sabon by Swales & Willis Ltd, Exeter, Devon

Contents

Preface and acknowledgements

This book describes the resistance of concrete to chloride ingress, and covers chloride transport mechanisms, test methods and prediction models. Owing to the worldwide problem of chloride-induced corrosion of the reinforcement in concrete structures, testing and modelling chloride ingress in concrete has been one of the hottest topics in relation to the durability of reinforced concrete structures. However, quite often chloride transport in concrete is inadequately represented because it involves very complicated mechanisms. Without explicit clarification of transport parameters, units and assumptions, confusion and obscurity can easily be introduced into both research publications and engineering applications. This book aims to address these issues by sharing a summary of my and my co-authors' personal research and the application of the research to practice.

At the beginning of 1990s, I joined Professor Lars-Olof Nilsson's team at Chalmers University of Technology in Gothenburg, Sweden, and started my doctorate research work on this topic, with a literature survey based on RILEM TC 116-PCD 'Permeability as a Criterion of its Durability' as the starting point. From the literature survey, I found that there was no theoretical relationship between the Coulomb test (AASHTO T 277) and the chloride penetration of concrete. With the help of a Chinese textbook about mathematical and physical functions, I derived an analytical solution to the differential equation of diffusion and migration (the Nernst–Plank equation), which became the basis of a rapid test method – the rapid chloride migration (RCM) test. As part of the development of this test method and the associated prediction models, we tried to clarify various parameters and their units, as well as the assumptions behind various other methods and associated models.

After my PhD studies, I continued the work on this topic at SP Swedish National Testing and Research Institute (later renamed the SP Technical Research Institute of Sweden). Together with the Chalmers group, we were involved in several research projects at national, Nordic and European levels, dealing with the durability of concrete in chloride-exposed environments. We were also involved in an international technical committee, RILEM TC-178 TMC 'Testing and Modelling Chloride Penetration in Concrete'. From the preliminary work by RILEM TC-178 the need for further research was identified, and this led to a European Union project entitled 'Resistance of Concrete to Chloride Ingress – From Laboratory Tests to In-field Performance' or, in short form, the CHLORTEST, coordinated by the SP Technical Research Institute of Sweden and involving 17 partners from ten European countries. Following this European project, a number of field investigations were conducted, from which more in-field performance data became available for the further validation of test methods and prediction models.

This book is my attempt to share my improved understanding of the mechanisms of chloride transport and the related models and test techniques. With the encouragement of Senior Editor Tony Moore at Taylor & Francis, I and my co-authors, together with Professor Muhammed Basheer at Queen's University Belfast, North Ireland, compiled this book, with the intention to present the state-of-the-art knowledge and understanding of chloride ingress into concrete to new researchers, engineers and practitioners, as well as owners and administrators of concrete structures.

Professor Lars-Olof Nilsson contributed most of the content of Chapters 4 and 5, and part of the content of Chapters 2 and 6. Professor Muhammed Basheer, together with his colleague Dr Sreejith Nanukuttan, contributed most of the content of Section 3.4, which expands the use of traditional laboratory tests to *in situ* applications. Professor Muhammed Basheer also contributed a part of the content of Chapters 1 and 6, and reviewed the whole book, making many constructive revisions, both technically and linguistically.

As the first author, I would express my deepest gratitude to the second author, Professor Lars-Olof Nilsson, who was my supervisor and torchbearer during my PhD studies, which enlightened my every step towards a deep knowledge and understanding of mass transport in porous building materials. It is my hope that, in turn, this book will function as a torch to light the way for more junior researchers to further research ideas, and as a tool or guide for engineers and practitioners with regard to their practical engineering applications, as well as to owners and administrators of concrete infrastructures to assist in their decision-making when planning maintenance and repairs in relation to the durability of concrete structures in chloride environments.

Finally, I would like to acknowledge all the partners in the EU project CHLORTEST for their contributions to the project work, which resulted in fruitful outcomes, and which also form part of this book. They are (excluding the SP Technical Research Institute of Sweden as coordinator):

- Drs Carmen Andrade and Malta Castellote, Institute of Construction Sciences 'Eduardo Torroja' (IETcc), Spain;
- Prof. Miguel A. Climent, University of Alicante (UoA), Spain;
- Dr Anders Lindvall, Chalmers University of Technology (Chalmers), Sweden;

- Mr Steinar Helland, Skanska Norge AS (Selmer), Norway;
- Mr Samir Redha, Swedish National Road Administration (SNRA), Sweden;
- Dr Ilie Petre-Lazars, Electricité de France (EDF), France;
- Prof. Rob Polder, Netherlands Organisation for Applied Scientific Research (TNO), Netherlands;
- Prof. Jörg Kropp, Hochschule Bremen (HSB), Germany;
- Dr Andraz Legat, Slovenian National Building and Civil Engineering Institute (ZAG), Slovenia;
- Prof. Muhammed Basheer and Dr. Sreejith Nanukuttan, Queens University Belfast (QUB), UK;
- Dr Manuela Salta, Laboratório Nacional de Engenharia Civil (LNEC), Portugal;
- Dr Gisli Gudmundsson, Icelandic Building Research Institute (IBRI), Iceland;
- Dr Myriam Carcasses, National Institute of Applied Science (INSA), France;
- Dr Véronique Baroghel-Bouny, Laboratoire Central des Ponts et Chaussées (LCPC), France;
- Ms Mª Carmen Naranjo Martinez, Valenciana de Cementos, S.A. CEMEX (VCLC), Spain;
- Prof. Lars-Olof Nilsson, Lund Institute of Technology (LTH), Sweden.

Without their contributions it would not have been possible to complete the project and obtain the fruitful results that form the basis of this book.

Tang Luping
Gothenburg, February 2011

Nomenclature

Unless otherwise specified in the text, the following notations and units are used in this thesis:

Symbol	Unit	Description and definition
a		Factor of electrical potential, $a = zFU/RTL$
a	m or cm	Spacing of the probes
A		Cross-sectional area
$A_{material}$		Cross-sectional area of the material
A_{pore}		Area of the pores on any cross-section of the material
$A_{solution}$		Area of the solution
B		Proportionality factor
c	$kg_{Cl}/m^3_{solution}$	Concentration of free chloride in general, by volume of solution
c_0	$kg_{Cl}/m^3_{solution}$	Concentration of free chloride in a bulk solution or an upstream diffusion cell
c_1	$kg_{Cl}/m^3_{solution}$	Concentration of free chloride in a downstream diffusion cell
c_b	$kg_{Cl}/m^3_{solution}$	Concentration of bound chloride by volume of solution
c_d	$kg_{Cl}/m^3_{solution}$	Concentration of free chloride at which the colour changes when using a colorimetric method to measure the chloride penetration depth
c_i	$kg_{Cl}/m^3_{solution}$	Initial concentration of free chloride in pore solution
c_s	$kg_{Cl}/m^3_{solution}$	Concentration of free chloride in a saturated solution
c_t	$kg_{Cl}/m^3_{solution}$	Concentration of total chloride by volume of solution
C_b	$kg_{Cl}/kg_{material}$	Content of bound chloride by weight of material

C_{bm}	$kg_{Cl}/kg_{material}$	Monolayer adsorption capacity of bound chloride
C_t	$kg_{Cl}/kg_{material}$	Content of total chloride by weight of material
C_{ti}	$kg_{Cl}/kg_{material}$	Initial content of total chloride in concrete
C_{ts}	$kg_{Cl}/kg_{material}$	Content of total chloride at the surface of concrete
COV		Coefficient of variation
D	m^2/s	Diffusion coefficient in general
D_0	m_x^2/s	Intrinsic diffusion coefficient
D_{nss}	m_x^2/s	Diffusion coefficient obtained from a non-steady-state diffusion test
D_{nssm}	m_x^2/s	Diffusion coefficient obtained from a non-steady-state migration test
D_{PD}	$\dfrac{m_{solution}^3 \cdot m_x}{m_{concrete}^2 \cdot s}$	Potential diffusion index obtained from a steady-state migration test
D_{ss}	$\dfrac{m_{solution}^3 \cdot m_x}{m_{concrete}^2 \cdot s}$	Diffusion coefficient obtained from a steady-state diffusion test
D_{ssm}	$\dfrac{m_{solution}^3 \cdot m_x}{m_{concrete}^2 \cdot s}$	Diffusion coefficient obtained from a steady-state migration test
E	V/m	Density of electrical field
E_b	J/mol	Activation energy for chloride binding
E_D	J/mol	Activation energy for chloride diffusivity
erf		Error function
erf^{-1}		Inverse of the error function
erfc		Error function complement, erfc = 1 − erf
f_b		Binding constant
f_c		Weight fraction of cement in the total solid material in the concrete mixture
f_{OH}		Hydroxide-dependent coefficient
F	$J/(V\ mol)$	Faraday constant, $9.648 \times 10^4\,J/(V\ mol)$
I_0	ampere	Current immediately after the voltage is applied in the Coulomb test
I_t	ampere	Current at t minutes after the voltage is applied in the Coulomb test
J	$kg/(m^2\ s)$	Flux of ions in general
J_0	$\dfrac{kg_{Cl}}{m_{solution}^2 \cdot s}$	Flux of ions through a unit area of solution
J_d	$kg/(m^2\ s)$	Diffusion flux of ions
J_m	$kg/(m^2\ s)$	Migration flux of ions

K_b	$m^2_{solution}/kg_{gel}$	Chloride binding factor in a non-steady-state migration test
L	m	Thickness of specimen
m_x		Meter in the transport direction
N	$\dfrac{1}{m^3_{solution}}$	Number density of ions
P_{sol}	$m^3_{solution}/m^3_{material}$	Porosity or content of water-accessible pores by volume of material
Q	coulomb	The charge passed in the Coulomb test
r		Correlation coefficient
R	J/(mol K)	Gas constant, 8.314 J/(mol K)
t	s	Time variable, age or test/exposure duration
t_i		Transference number of the ith ion
T	K	Absolute temperature
u	$m^2/(s\ V)$	Ion mobility
U	V	Electrical potential difference
v	V or mV	Measured electrical potential difference
$V_{material}$		Volume of the material
V_p	$m^3_{solution}/kg_{material}$	Content of the pore solution, or water-accessible pores, in the material
W	$kg_{material}$	Weight of dry material
W_n	$kg_{water}/m^3_{material}$	Content of non-evaporable water by volume of material
W_n^0	kg_{water}/kg_{cement}	Specific non-evaporable water when cement is fully hydrated
w/c or w/c		Water/cement ratio
x	m	Distance variable
x_d	m	Depth of chloride penetration measured using a colorimetric method
x_f	m	Inflection point of a free chloride profile in a non-steady-state process of diffusion–migration
z		Ion valence
α		Constant in general
α_h		Degree of hydration
α_{OH}		Constant for hydroxide-dependent coefficient
β		Constant in general
ϕ	V	Electrical potential
γ		Activity coefficient
γ_b		Constant for chloride binding in a non-steady-state migration test

ε	$m^3_{solution}/m^3_{material}$	Porosity or content of water-accessible pores by volume of material
μ	J/mol	Chemical potential
$μ_0$	J/mol	Standard chemical potential (constant)
ν	m/s	Average diffusion velocity of ions
$ν_m$	m/s	Average migration velocity of ions
$ρ_{Wenner}$	Ω·m or kΩcm	Surface resistivity of a Wenner probe
σ		Conductivity of concrete
$λ_i$		Molar conductivity of the ith ion
$Λ_m$	$Ω^{-1} m^2 mol^{-1}$	Molar conductivity

Indexes

0	Reference point in general
b	Binding; bound
c	Cement
d	Depth; diffusion
D	Diffusivity
h	Hydration
i	Index of distance
i or ini	Initial
j	Index of time
m	Migration
n	Non-evaporable
nss	Non-steady-state
nssm	Non-steady-state migration
p	Pores; pore solution
s	Saturation; surface; surface zone
ss	Steady-state
ssm	Steady-state migration
t	Time; age
T	Temperature
x	Depth
x	x coordinate

Abbreviations and acronyms

AASHTO	American Association of State Highway and Transportation Officials, Washington, DC, USA
AEC	A previous Danish engineer consulting company
ASTM	ASTM International, originally known as the American Society for Testing and Materials, West Conshohocken, PA, USA

CEN	European Committee for Standardization, Brussels, Belgium
Chalmers (or CTH)	Chalmers University of Technology, Gothenburg, Sweden
CHLORTEST (ChlorTest)	EU research project, 'Resistance of Concrete to Chloride Ingress – From Laboratory Tests to In-Field Performance', under the 5th FP GROWTH, 2003–2005
ClinConc	A mechanistic model based on free chloride diffusion and non-linear binding for the prediction of chloride ingress in concrete, developed at Chalmers, Gothenburg, Sweden
ConLife	EU research project, 'Life-time Prediction of High-Performance Concrete with Respect to Durability', under the 5th FP GROWTH, 2002–2004
DuraCrete	EU research project, 'Probabilistic Performance based Durability Design of Concrete Structures', under Brite EuRam III, 1996–1999
EDF	Electricité de France
Elkem	Manufacturer of silica fume (micro-silica), Kristiansand, Norway
erfc	Error function complement
EU	European Union
fib	International Federation for Structural Concrete, Lausanne, Switzerland
FL	Fly ash
FORCE	A technological service and research organisation in Brøndby, Denmark
FORM	First-order reliability method
HETEK	Danish research project on high performance concrete, 1995–1997
IETcc	Institute of Construction Sciences 'Eduardo Torroja', Madrid, Spain
INSA	National Institute of Applied Science, Toulouse, France
ISO	International Organisation for Standardisation, Geneva, Switzerland
LEO model	An empirical model for the prediction of chloride ingress, developed at EDF, Paris, France
LNEC	Laboratório Nacional de Engenharia Civil, Lisbon, Portugal

MsDiff	A mechanistic model based on multi-species diffusion, for the prediction of chloride ingress; developed at INSA, Toulouse, France
Nordtest	An inter-governmental organisation for standardised testing in the Nordic countries
NTNU	Norwegian University of Technology and Science, Trondheim, Norway
OPC	Ordinary Portland cement
PC	Portland cement
RCM	A test method based on the electrochemical principle for rapid chloride migration
RILEM	International Union of Laboratories and Experts in Construction Materials, Systems and Structures, Paris, France
RWS	Ministry of Transport, Public Works and Water Management of the Netherlands
Selmer-Skanska	Construction company, Oslo, Norway
SF	Silica fume
SINTEF	A research organisation in Trondheim, Norway
SL	Slag
SORM	Second-order reliability method
SP	Swedish National Testing and Research Institute (currently Technical Research Institute of Sweden), Borås, Sweden
SRPC	Sulphate-resistant Portland cement
Teknologisk	Danish Technology Institute, Taastrup, Denmark
TNO	Netherlands Organisation for Applied Scientific Research, Delft, The Netherlands

1 Introduction

Civil infrastructure is the backbone of any society. When the economy of a country is strong, its investment in infrastructure projects goes up, and vice versa, because progress in a society is interlinked with infrastructure projects. It is normally considered that most developed countries spent more than 50% of their infrastructure budget on the restoration of structures which suffer from premature deterioration. This is for a number of reasons, the principal one of which is chloride-induced corrosion of steel in reinforced concrete structures. Thus, if more investment is to be made in new infrastructure projects, there is a need to reduce the money spent on repair and rehabilitation of existing structures because of chloride-induced corrosion of reinforcing steel.

Chloride itself does not, in normal circumstances, result directly in any damage to concrete, but it can induce corrosion of steel within concrete. This has very important economic and social consequences, due to the need to divert funds to the repair of damaged areas and sometimes to close the facility for the repair and rehabilitation work. To ensure the quality of expensive concrete structures and to properly maintain them during the designed service life, a good understanding of the mechanisms of chloride transport, test methods and relevant models for predicting the service life are needed.

There are many sources of chlorides, and these are summarised well by Poulsen and Mejlbro (2006) in their book *Diffusion of Chloride in Concrete: Theory and Application*. Amongst these sources, seawater in marine environments and de-icing salts used in the winter on roads to melt the ice are two very common sources of chlorides with regard to concrete structures. The effect of chlorides on reinforced concrete structures can be severe corrosion, with extensive delamination and spalling of concrete, as shown in Figure 1.1, or pitting corrosion, with staining of concrete surface, as shown in Figure 1.2. Both the bridges in Figures 1.1 and 1.2 were on the same road network, but due to differences in material properties and construction practices the nature of deterioration mechanism is different. This means that not only the mechanisms of chloride transport need to be thoroughly understood, but methods of assessing the resistance of concretes to the chloride transport and associated chloride transport models are required for a

Figure 1.1 Salt water run down on a bridge abutment, causing chloride-induced corrosion of steel and extensive spalling of concrete.

Figure 1.2 Extensive rust staining on a bridge abutment caused by chloride-induced corrosion of steel.

reliable prediction of the service life of concrete structures in chloride exposure environments. Furthermore, there is a need to verify the validity of tests and chloride transport models by applying to structures in service.

Reinforcing steel in a concrete structure is normally protected by the cover of concrete, which keeps the steel in alkaline passive conditions and resists the ingress of aggressive substances, such as chlorides, sulphates and carbon dioxides. Chlorides must 'penetrate' into concrete and reach the steel in sufficient quantity in order to depassivate the steel and induce corrosion. Clearly, this means that a study of chloride ingress into concrete is the first step in being able to predict service life of reinforced concrete structures exposed to chloride environments. Therefore, since the 1970s, the topic of chloride ingress into concrete has attracted the attention of many researchers, engineers, practitioners, and owners and administrators of expensive reinforced concrete infrastructures.

1.1 Challenges when predicting the service life of reinforced concrete structures

Chloride ingress into concrete involves complex physical and chemical processes. The complexity comes from at least three sources:

1   The external environment is not constant. In marine environments, the amount of chlorides in contact with concrete depends on whether the structure is fully submerged or in the tidal zone, or in contact only with marine fog; while for roads, the intermittent use of de-icing salts in a very cold climate makes it difficult to calculate the amount of chlorides sprayed onto the structures.
2   Concrete is composed of different types of cement and binder, in different mix proportions, and thus is not a single material, being different in different structures. Furthermore, the properties of the hydrated cementitious materials evolve with age.
3   The mechanisms of chloride penetration are not confined to one transport process (such as diffusion), and may be a combination of convection (absorption), chemical and physical binding (adsorption), and interaction with coexisting ions. Changes in temperature, rain and sunshine introduce variations that should also be taken into account.

Owing to its important role with regard to the durability of concrete structures, many methods and models have been proposed for testing and predicting chloride ingress in concrete. However, until now the complexity of the process has prevented the reaching of a general agreement on a single test method or prediction model.

1.2 Progress in test methods for measuring chloride transport

In North America, by the 1980s two methods for testing the resistance of concrete to chloride ingress had been developed and standardised:

1   AASHTO T259 (since 1980), *Standard Method of Test for Resistance of Concrete to Chloride Ion Penetration*, and later as ASTM C1543

(since 2002), *Standard Test Method for Determining the Penetration of Chloride Ion into Concrete by Ponding*;

2   AASHTO T277 (since 1983), *Standard Method of Test for Rapid Determination of the Chloride Permeability of Concrete*, and later as ASTM C 1202 (since 1994), *Standard Test Method for Electrical Indication of Concrete's Ability to Resist Chloride Ion Penetration* (this title has also been adopted in the latest version of AASHTO T277).

The first method, namely the ponding test, is laborious and time consuming, the test taking longer than 90 days. The second method, namely the rapid chloride permeability test (RCPT), is an indirect test based on electrical resistivity, and is a pioneer amongst accelerated tests. From the results of these two tests it is difficult to derive clear chloride transport properties in concrete. Therefore, the two methods are basically used for comparison or rough ranking of the ability of concrete to resist the ingress of chlorides.

Following the North American experience, three test methods were developed and standardised in Nordic countries:

1   NT BUILD 355 (since 1989), *Concrete, Mortar and Cement Based Repair Materials: Chloride Diffusion Coefficient from Migration Cell Experiments*, based on the principle of steady-state migration;

2   NT BUILD 443 (since 1995), *Concrete, Hardened: Accelerated Chloride Penetration*, based on the principle of non-steady-state diffusion under high chloride concentration; and

3   NT BUILD 492 (since 1999), *Concrete, Mortar and Cement Based Repair Materials: Chloride Migration Coefficient from Non-Steady State Migration Experiments*, based on the principle of non-steady-state migration.

From each of these three test methods, transport properties such as diffusion or migration coefficient can be derived. Therefore, the last two of these Nordic methods were later adopted, with certain modifications, in North America, as:

1   AASHTO PT 64 (since 2003), *Standard Method of Test for Predicting Chloride Penetration of Hydraulic Cement Concrete by the Rapid Migration Procedure*; and

2   ASTM C1556 (since 2004), *Standard Test Method for Determining the Apparent Chloride Diffusion Coefficient of Cementitious Mixtures by Bulk Diffusion*.

At the European level, the standardisation processes are slow, owing to geographical and climatic differences across the continent, as well as possibly political and philosophical differences. Not until 2004 was an immersion test at an elevated temperature (40°C) standardised:

• EN 13396 (since 2004), *Products and Systems for the Protection and Repair of Concrete Structures – Test Methods – Measurement of Chloride Ion Ingress.*

This method is basically similar to the American ponding test (AASHTO T259 or ASTM C1543), and is used to compare the chloride ingress in repair materials at a few penetration depths that of a control specimen after immersion of up to 6 months.

Besides the above-mentioned standardised methods, there are a number of other methods, which are variations of the above-mentioned ones. Unfortunately, the results obtained using different test methods are, in general, not directly comparable, due to the complex mechanisms involved in chloride transport in concrete.

1.3 Mathematical models for describing chloride transport

Chloride transport models are normally used either to describe the advance of the chloride profile through concrete, or to predict the time for the chloride front to reach a certain depth (normally the depth of the reinforcement) and at a certain concentration (normally the chloride threshold value for the initiation of corrosion). The latter enables the prediction of the service life of reinforced concrete structures based on the initiation of chloride-induced corrosion. Traditionally, chloride transport has been expressed by means of Fick's laws of diffusion. The first law is used when steady-state conditions exist, and the second law is used when the transport of chlorides occurs under non-steady-state conditions. Most service-life models published towards the end of the twentieth century made use of one of these two laws of diffusion. However, it is now realised that chloride transport is a complex subject, and hence not only should the transport of chlorides by diffusion be given emphasis, but also the interactions of the chlorides with other ions and the medium itself are important in describing fully the chloride transport. This is particularly the case when cements containing supplementary cementitious materials are used as binders.

1.4 Prediction of the service life of structures in chloride-exposed environments

In the past decade, the parameter known as the 'chloride diffusion coefficient' has been used in service-life design or durability specifications in numerous large construction projects, such as the Great Belt Bridge in Denmark (completed in 1998), the Øresund Bridge and Tunnel between Denmark and Sweden (completed in 1999), the Vasco da Gama Bridge in Lisbon, Portugal (completed in 1998), the Donghai Bridge in Shanghai, China (completed in 2005), and the Hangzhou Bay Bridge in Hangzhou, China (the world longest cross-sea bridge with the total length of 35.7 km, completed in 2007).

However, some premature failures and discussions at congresses have indicated that this design tool is far from being accurate. Regardless of this, the first step is to present evidence on the relationship between a short-term test in the laboratory and the long-term performance in reality. With this purpose, the EU project *Resistance of Concrete to Chloride Ingress – From Laboratory Tests to In-field Performance (CHLORTEST)* was funded by the European Community under the Competitive and Sustainable Growth Programme in the 5th Framework in the middle of 2003–2005. The project was coordinated by SP Swedish National Testing and Research Institute (currently SP Technical Research Institute of Sweden). The project consortium consisted of 17 partners from ten European countries, and included testing and research centres/institutions, universities, material suppliers, construction contractors and owners. The project objectives were:

- to evaluate different laboratory performance test methods in terms of the theoretical basis, technical feasibility, measurement precision and applicability in practical construction design and quality assessment;
- to recommend two reliable methods, one reference method and another rapid method, for testing the resistance of concrete to chloride ingress;
- to collect in-field performance data on chloride ingress and reinforcement corrosion;
- to verify the laboratory performance tests with the collected in-field performance data using different models, including scientific, empirical and probabilistic approaches;
- to recommend the practical use of the laboratory performance tests and the interpretation of the test results, including a proposal for acceptance criteria.

In addition to this large European cooperative research project, a number of Swedish research projects were conducted with the purpose of investigating the long-term in-field performance and developing rapid laboratory test methods and prediction models. Two field exposure sites, one in a marine environment and the other in a de-icing salt environment, were established in order to collect the in-field data from the concrete specimens, which were exposed under relatively controlled conditions. These data have been proven very valuable in the verification of short-term laboratory test methods and various prediction models.

1.5   The structure of this book

This book presents a summary of the results of the projects mentioned in the preceding sections, and results from other countries, and the aim is to present the state-of-the-art knowledge and understanding of chloride transport mechanisms, test methods and prediction models. The book consists of five chapters in addition to this introductory chapter.

In Chapter 2 the basic mechanisms and theories behind chloride transport in concrete are described. The commonly used parameters and units for expressing chloride ions and their transport properties in concrete are classified. Various mechanisms involved in chloride ingress into concrete are reviewed and discussed.

Test methods applicable in the laboratory and the field are described and discussed in Chapter 3, along with results on the precision of the methods obtained inter-laboratory comparison tests.

In Chapter 4 some of the fundamentals of the various chloride ingress models are explained. The different types of model are analysed theoretically and critically. The sensitivities of the various models are analysed by means of both analytical and probabilistic approaches.

In Chapter 5 the results of a benchmarking evaluation of the different models are presented and discussed. Some results of validation studies of these models against long-term exposure data taken from both the field exposure sites and real concrete structures are presented.

Finally, guidelines for the practical use of the test methods and models are given in Chapter 6, and the test methods proposed, based on the precision results, are described in an Appendix.

2 Chloride transport in concrete

2.1 Introduction

As mentioned in Chapter 1, chloride transport in concrete is a rather complicated process, which involves ion diffusion, capillary suction and convective flow with flowing water, accompanied by physical and chemical binding. Sometimes, an external electrical potential may be imposed on a concrete for the removal of chlorides or for a rapid determination of chloride diffusivity. In this case, the mechanism of migration is involved. In this chapter the commonly used parameters, and their units, for expressing chloride concentrations and related transport properties in concrete are introduced, and the current understanding of the two main mechanisms of chloride ion transport – ion diffusion and migration – is summarised. Other mechanisms, such as absorption, convection, permeability and wick action, may also contribute to chloride transport, and hence these are also briefly discussed.

2.2 Chloride concentrations

Chlorides exist in concrete in two forms: free and bound. It is the general understanding that only free chloride ions in the pore solution in concrete are movable. However, in practice it is difficult to separate the free chlorides from the total chlorides (free plus bound). Both the free and the total chlorides are often determined using classical chemical analysis. The amount of chloride ions in concrete may, therefore, be expressed in a number of ways. Nilsson *et al.* (1996) have summarised the most frequently used expressions, as follows:

$$c_I \quad (\text{kg Cl/m}^3 \text{ of solution}) \qquad\qquad (2.1\!:\!1)$$

$$c_{II} \quad (\text{kg Cl/m}^3 \text{ of pore solution}) \qquad\qquad (2.1\!:\!2)$$

$$c_{III} \quad (\text{kg Cl/m}^3 \text{ of material}) \qquad\qquad (2.1\!:\!3)$$

$$c_{IV} \quad (\text{kg Cl/m}^3 \text{ of solid material}) \qquad\qquad (2.1\!:\!4)$$

c_V (kg Cl/kg of cement or binder)   (2.1:5)

c_{VI} (kg Cl/kg of gel ('gel' = hydrated cement + bound water))   (2.1:6)

c_{VII} (kg Cl/kg of concrete)   (2.1:7)

c_{VIII} (kg Cl/kg of sample)   (2.1:8)

Consequently, giving the quantity of chlorides in only kg/m³, kg/kg or % might cause confusion and large errors when trying to use the data. The unit of kilogram (kg) for mass is sometimes replaced by moles (mol). It does not change the relations and the relation between kg Cl and mol Cl is obvious. Concentrations in g/l are identical to concentrations in kg/m³ of solution, but it should be clearly stated whether the mass is that of chloride ions or salt (NaCl); for instance, g Cl per litre or g NaCl per litre. The difference in quantity between chloride ions and NaCl is greater than 50%, due to the different molar weights (35.45 g Cl and 58.45 g NaCl).

Most of the relations between the different concentrations in Eq. 2.1:1 to Eq. 2.1:8 are simple. The concentrations c_{III} to c_{VIII} are related to each other by the dry density of the material, dry density of the solid material, cement content, degree of hydration, the moisture content and the molar weight of chloride.

The relation between c_{II} and c_{III} to c_{VIII} is more difficult, however:

$$c_{II} = c_{III}/p_{sol}$$ (2.1:9)

where the problem is to define the porosity p_{sol}. The term p_{sol} includes only that part of the porosity which contains a liquid that acts as a solvent. One might question whether all the pore water really acts as a solvent, i.e. p_{sol} = total porosity. Measurements made by Mangat and Molloy (1994) give some information. They measured the concentration of free chlorides. When multiplied by the total amount of pore water, this gave more free chlorides than the total that was actually present in the samples. If the measured concentrations are correct, it is obvious that not all the pore water acts as a solvent. Thus further research is needed to identify the relationship between total chlorides, total amount of pore water and free chlorides. Until this has been done, the capillary porosity or the empty porosity at a low humidity, such as 11% or 45%, may be used to determine the free chlorides.

2.3 Chloride binding/interaction and binding capacity

When free chloride ions from environmental solutions penetrate into concrete, some of them will be captured by the cement hydration products. This is called 'chloride binding'. It is generally believed that both physical adsorption and chemical reaction are involved in chloride binding. The huge surface area of hydration gel supplies plenty of sites for the physical binding, while the formation of 'Friedel's salt' is commonly thought to be a major product of chemical binding with hydration products. The bound chlorides are generally thought to be 'harmless' to the reinforcement. However, as the physically bound chlorides can be released or desorbed (Tang and Nilsson, 1991), due to mechanisms such as carbonation or sulphate ingress, there is an argument against the harmlessness of bound chlorides. However, despite this argument, the binding effect can retard the transport of the free chloride ions. Therefore, the effect of chloride binding needs to be taken into account when studying chloride transport in concrete.

The mechanisms of chloride binding or interactions between chloride and the matrix of cement-based materials are not very well understood. Instead, for most applications a 'binding isotherm' is used to give the relation between the free and the bound chlorides. In the literature, chloride binding isotherms have been mathematically expressed in different ways.

2.3.1 Linear chloride binding

Tuutti (1982) suggested a simple linear relationship between the bound chloride content C_b and the free chloride concentration c:

$$C_b = \alpha \cdot c \tag{2.2}$$

where α is a constant. Note that the capital letter C denotes content by weight or mass of material, while the lower case c denotes concentration by volume of solution. Unfortunately, the experimental data reported by other researchers, e.g. Ramachandran *et al.* (1984), Andrade and Page (1986) and Arya *et al.* (1987), do not fit the above oversimplified equation. Arya and Newman (1990) proposed a linear relationship with an intercept on the axis for both bound and total chlorides:

$$C_b = \alpha \cdot c + \beta \tag{2.3}$$

where β is a constant (intercept). The coefficients α and β will be different for bound and total chlorides. Although Eq. 2.3 fitted their experimental data fairly well, it cannot explain the physical meaning of $C_b = \beta$ at $c = 0$, especially when considering external chloride penetration. For instance, if a chloride-free concrete is exposed to a chloride-free solution, bound or total chloride should be zero, not β. Therefore, Eq. 2.3 is not applicable at low chloride concentrations, at which the experimental data diverge remarkably from linearity, as pointed out by Mangat and Molloy (1994).

2.3.2 Non-linear chloride binding

In fact, the linear relationship holds only in a limited range of free chloride concentrations. In most cases the relationship between bound and free

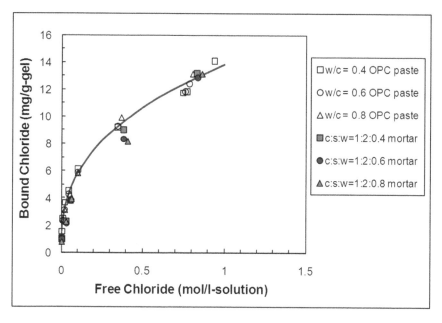

Figure 2.1 Example of non-linear chloride binding. c, cement; s, sand; OPC, ordinary Portland cement; w, water. (Based on Tang and Nilsson (1993b).)

chlorides is non-linear, according to the experimental data reported by, for instance, Richartz (1969), Theissing *et al.* (1978), Pereira and Hegedus (1984), Blunk *et al.* (1986), Tritthart (1989a, 1989b), Byfors (1990), Sandberg and Larsson (1993), Tang and Nilsson (1993b), Bigas (1994), Akita and Fujiwara (1995) and Delagrave *et al.* (1996). An example of non-linear chloride binding is shown in Figure 2.1.

2.3.3 Chloride-binding capacity

The *binding capacity* is the capacity of a material to bind chlorides when the ion concentration changes, and is defined as:

$$\text{Binding capacity} = \frac{\partial c_b}{\partial c} \tag{2.4}$$

where c_b is the bound chloride concentration (as expressed in the solution, because the actual bound chlorides are distributed in the solid phase). This is the slope of a 'binding isotherm' with linear scales, as shown in Figure 2.1. The dimension of the binding capacity depends on the units chosen for c_b and c, respectively. Some examples are shown in Figure 2.2. Obviously, the binding capacity depends on the concentration, as seen from Figure 2.2, but

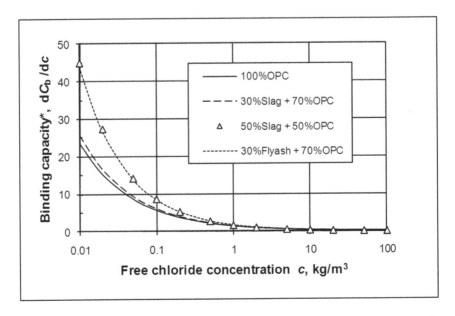

Figure 2.2 Examples of chloride-binding capacity (see Tang (1996b)). OPC, ordinary Portland cement.

in some models, especially empirical models, it is assumed to be constant, as discussed in Chapter 4.

One important, but not obvious, question is whether the concentration of chloride in the pore water is equal to the concentration in the surrounding seawater or exposure solution. Some recent procedures for determining the 'free' chloride by leaching into water cause confusion about 'free' and 'water-soluble' chlorides. It should be remembered that the quantity of 'water-soluble' chloride is strongly dependent on the test method used. Different test methods can result in different amounts of 'water-soluble' chlorides, which is not necessarily equal to 'free' chlorides in the pore solution.

Chloride binding is, from measurements, a function of several parameters: concentration of free chloride, pH, temperature, moisture content, gel content, type of binder, binder content, water/binder ratio, degree of hydration, duration of exposure, etc. An example of the pH dependency is shown in Figures 2.3 and 2.4. The mechanisms behind all these observations are not fully understood, and much more research is needed and much better measurements must be performed (see also Larsen (1998)).

2.4 Ion diffusion

Diffusion is the movement of a substance under a gradient of concentration or, more strictly speaking, chemical potential, from an area of high

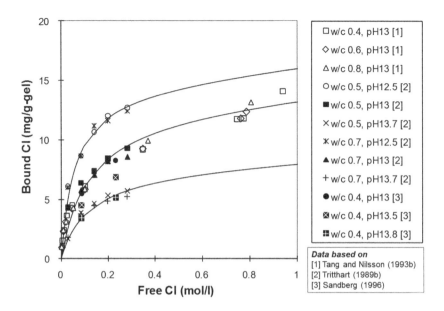

Figure 2.3 Effect of pH on chloride binding (see Tang and Nilsson (1995)). c, cement; w, water. Data based on: [1] Tang and Nilsson (1993b); [2] Tritthart (1989b); [3] Sandberg (1996).

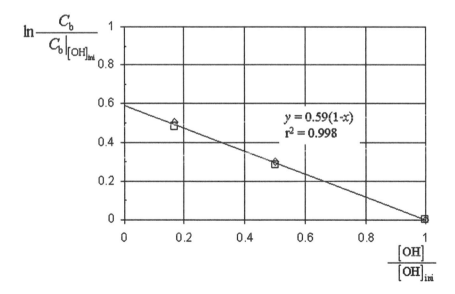

Figure 2.4 An example of the pH dependency of the chloride binding isotherm (see Tang (1996b)).

concentration to an area of low concentration. When discussing chloride transport, it should be remembered that only free chloride ions in a solution could contribute to a concentration or chemical potential.

2.4.1 Diffusion function and Fick's first law

Consider a liquid system, as shown in Figure 2.5, in which there are diffusants, i.e. chloride ions. The chemical potential of the chloride ions is:

$$\mu = \mu_0 + RT \ln (\gamma c) \tag{2.5}$$

In the presence of a chemical potential gradient, the ions will move along the gradient. As chloride ions (being anions) cannot exist alone, there must be some cations around. Once the chloride ions move forward, a counter electrical field ϕ' between the chloride ions and the surrounding cations may be formed. This counter electrical field tends to draw back the chloride ions. Therefore, the average velocity of chloride ion movement v is

$$v = -B\frac{\partial\mu}{\partial x} + u'\frac{\partial\phi'}{\partial x} = -BRT\left(\frac{\partial\ln c}{\partial x} + \frac{\partial\ln\gamma}{\partial x}\right) + u'\frac{\partial\phi'}{\partial x} \tag{2.6}$$

or

$$v = -BRT\frac{\partial\ln c}{\partial x}\left(1 + \frac{\partial\ln\gamma}{\partial\ln c}\right) + u'\frac{\partial\phi'}{\partial x} \tag{2.6:1}$$

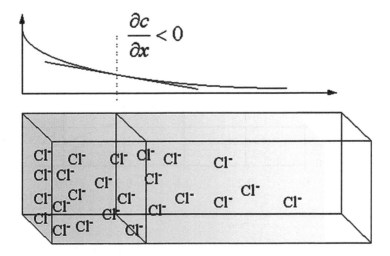

Figure 2.5 Illustration of chloride diffusion in a pure solution system.

where B is the proportionality factor, R is the gas constant, T is the absolute temperature and u' is the ion mobility.

Putting $D = BRT$, where D is the diffusion coefficient, the flow rate of the ions through a unit area of solution, called the flux, is:

$$J_d = c \cdot v = -D \frac{c \partial \ln c}{\partial x}\left(1 + \frac{\partial \ln \gamma}{\partial \ln c}\right) + cu' \frac{\partial \phi'}{\partial x}$$

where J_d denotes the diffusion flux of ions and x is the distance travelled, or

$$J_d = -D \frac{\partial c}{\partial x}\left(1 + \frac{c}{\gamma} \cdot \frac{\partial \gamma}{\partial c}\right) + cu' \frac{\partial \phi'}{\partial x} \tag{2.7}$$

Both the activity coefficient γ and the counter electrical field ϕ' are dependent on the free chloride concentration and the other ions coexisting in the solution. It is very difficult to solve for the terms

$$\frac{c}{\gamma} \cdot \frac{\partial \gamma}{\partial c} \quad \text{and} \quad cu' \frac{\partial \phi'}{\partial x}$$

although some attempts have been made (see Chaterji (1994) and Zhang and Gjørv (1996)). These terms are, therefore, often neglected. Thus Eq. 2.7 becomes a general form of Fick's law, or Fick's first law:

$$J_d = -D \frac{\partial c}{\partial x} \tag{2.8}$$

It can be seen that the main assumptions in Eq. 2.8 are that $\partial \gamma = 0$ and $\partial \phi'/\partial x = 0$.

These assumptions are, of course, highly questionable, because under these assumptions the chloride ions are treated as uncharged 'particles'.

2.4.2 Steady-state diffusion and dimensions of the diffusion coefficient

It is well known that the dimension of the chloride diffusion coefficient D is m²/s. However, it is not clear to what m² is related. In the literature, few give a clear meaning of the diffusion coefficient. Nilsson (1992, 1993) was perhaps the first to identify that the different diffusion coefficients are confusing, and he classified different chloride concentrations and their relations to the diffusion coefficients.

It is apparent that, for a pure solution system, the flux J in Eq. 2.8 has the dimensions

$$\frac{kg_{Cl}/s}{m^2_{solution}}$$

and the concentration gradient $\partial c/\partial x$ has the dimensions

$$\frac{kg_{Cl}/m^3_{solution}}{m_{X\,coordinate}}$$

The diffusion coefficient D should, therefore, have the dimensions:

$$D = \frac{J_d}{-\dfrac{\partial c}{\partial x}} \;\Rightarrow\; \frac{\dfrac{kg_{Cl}/s}{m^2_{solution}}}{\dfrac{kg_{Cl}/m^3_{solution}}{m_x}} = \frac{\dfrac{kg_{Cl}/s}{m^2_{solution}}}{\dfrac{kg_{Cl}/\left(m^2_{solution}\cdot m_x\right)}{m_x}} = m^2_x/s \qquad (2.8{:}1)$$

In most cases, for chloride ions diffusing into concrete the source of chlorides is the solution outside the concrete, as shown in Figure 2.6. In the case of steady-state diffusion, the flux remains constant. This flux is often referred to as the 'flow rate through a unit area of concrete'. The diffusion coefficient obtained from the steady-state test, denoted by D_{ss}, becomes:

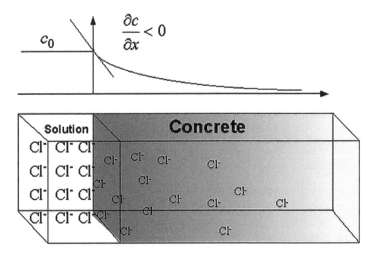

Figure 2.6 Illustration of chloride diffusion in a solution–concrete system.

$$D_{ss} = \frac{J_{concrete}}{-\dfrac{\partial c}{\partial x}} \Rightarrow \frac{\dfrac{kg_{Cl}/s}{m^2_{concrete}}}{\dfrac{kg_{Cl}/m^3_{solution}}{m_x}} = \frac{m^3_{solution} \cdot m_x}{m^2_{concrete} \cdot s} \tag{2.8:2}$$

It can be seen that the dimension of D_{ss} is complicated. In order to simplify the above dimension, consider a homogeneous pore structure, i.e. one in which the pore fraction in any thin slice along the x coordinate is equal to the porosity ε. This implies that the area fraction of pores on any cross-sectional surface is independent of the x coordinate. Thus the volume fraction of pores, or porosity, can be expressed as:

$$\varepsilon = \frac{V_{pore}}{V_{material}} = \frac{A_{pore}\int dx}{A_{material}\int dx} \Rightarrow \frac{m^3_{pore}}{m^3_{material}} = \frac{m^2_{pore} \cdot m_x}{m^2_{material} \cdot m_x} = \frac{m^2_{pore}}{m^2_{material}} \tag{2.9}$$

where V_{pore} is the volume of the pores in the material, $V_{material}$ is the volume of the material, A_{pore} is the area of the pores on any cross-section of the material, and $A_{material}$ is the cross-sectional area of the material. Under the saturated condition,

$$\frac{A_{solution}}{A_{material}} = \frac{A_{pore}}{A_{material}} = \varepsilon = p_{sol} \Rightarrow \frac{m^2_{solution}}{m^2_{material}} = \frac{m^2_{solution} \cdot m_x}{m^2_{material} \cdot m_x} = \frac{m^3_{solution}}{m^3_{material}} \tag{2.9:1}$$

where $A_{solution}$ is the area of solution, p_{sol} is the solution-filled porosity and m_x is the meter in the transport direction. Therefore,

$$\frac{D_{ss}}{p_{sol}} = \frac{1}{p_{sol}} \cdot \frac{J_{concrete}}{-\dfrac{\partial c}{\partial x}} \Rightarrow \frac{1}{\left(\dfrac{m^2_{solution}}{m^2_{concrete}}\right)} \cdot \frac{\left(m^2_{solution} \cdot m_x\right) \cdot m_x}{m^2_{concrete} \cdot s} = m^2_x/s \tag{2.10}$$

The coefficient having such a dimension can be defined as the 'intrinsic diffusion coefficient', denoted by D_0, which characterises the diffusion rate of chloride ions through the pore solution in concrete:

$$D_0 = \frac{D_{ss}}{p_{sol}} \tag{2.11}$$

To avoid the confusion of different dimensions, it is better to express the diffusion equation as:

$$J_{0d} = -D_0 \frac{\partial c}{\partial x} \tag{2.12}$$

where J_{0d} is the diffusion flow rate through a unit area of pore solution.

2.4.3 Non-steady-state diffusion

As mentioned in Section 2.2, chloride binding in concrete is inevitable. For a non-steady-state process, therefore, it is the increment in total chlorides that results in a difference in the flux, i.e.:

$$\frac{\partial c_t}{\partial t} = \frac{\partial c}{\partial t} + \frac{\partial c_b}{\partial t} = \frac{\partial c}{\partial t}\left(1 + \frac{\partial c_b}{\partial c}\right) = -\frac{\partial J_{0d}}{\partial x} = \frac{\partial}{\partial x}\left(D_0 \frac{\partial c}{\partial x}\right) \tag{2.13}$$

or, in the form of Fick's second law,

$$\frac{\partial c}{\partial t} = \frac{\partial}{\partial x}\left[\frac{D_0}{1 + \frac{\partial c_b}{\partial c}} \cdot \frac{\partial c}{\partial x}\right] = \frac{\partial}{\partial x}\left(D_{nss}\frac{\partial c}{\partial x}\right) \tag{2.14}$$

where D_{nss} denotes the diffusion coefficient obtained from a non-steady-state diffusion test:

$$D_{nss} = \frac{D_{ss}}{P_{sol}\left(1 + \frac{\partial c_b}{\partial c}\right)} \qquad m_x^2/s \tag{2.15}$$

Equation 2.15 is the relationship between the diffusion coefficients obtained from a steady-state test and a non-steady-state diffusion test (see Nilsson (1992) and Nilsson *et al.* (1994). Under the assumption of a constant D_{nss}, i.e. a constant chloride-binding capacity, Eq. 2.14 is simplified as a common expression of Fick's second law:

$$\frac{\partial c}{\partial t} = D_{nss}\frac{\partial^2 c}{\partial x^2} \tag{2.16}$$

As the chloride binding capacity $\partial c_b/\partial c$ changes greatly with the free chloride concentration c, as shown in Figure 2.2, the diffusion coefficient D_{nss} should not be constant, but is dependent on the free chloride concentration. A constant D_{nss} is based on a simplified model, which is discussed in Chapter 4.

Owing to the non-linear behaviour of chloride binding, it seems impossible to get an analytical solution to Eq. 2.16. Therefore, numerical approaches must be considered when dealing with the non-steady-state process of chloride transport in concrete.

2.5 Migration

Migration is the movement of a charged substance under the action of an electrical field. As in diffusion, only free chloride ions in a solution can contribute to the flow of migration.

2.5.1 Migration function

Under a gradient of electrical potential, the ions will move along a certain direction of the gradient, depending on the valence of the ions, at an average velocity v_m:

$$v_m = u \frac{\partial \phi}{\partial x} \tag{2.17}$$

where u is the ion mobility. According to the Einstein relation,

$$u = D \cdot \frac{zF}{RT} \tag{2.18}$$

where z is the ion valence and F is the Faraday constant.

It should be noted that Eq. 2.17 is valid for all ions in the solution. If the number density of chloride ions in a solution is N, in a time interval Δt the number of chloride ions passing through an area A should be $(A \cdot v_m \cdot \Delta t) \cdot N$. The number density N is another expression of concentration for a strong electrolyte such as NaCl solution. Therefore, the flux of chloride ions under the action of an electrical field should be:

$$J_m = \frac{(A \cdot v_m \cdot \Delta t) \cdot N}{A \cdot \Delta t} = N \cdot v_m \qquad \frac{\text{number of Cl ions}}{m^2_{solution} \cdot s}$$

or

$$J_m = c \cdot v_m = c \cdot D \cdot \frac{zF}{RT} \cdot \frac{\partial \phi}{\partial x} \qquad \frac{kg_{Cl}}{m^2_{solution} \cdot s} \tag{2.19}$$

Assuming a constant external electrical field, i.e. $\partial\phi/\partial x = U/L = E$, Eq. 2.19 becomes:

$$J_m = D \cdot \frac{zFE}{RT} \cdot c \tag{2.20}$$

The above equation is for pure migration. It can be seen from the above derivation that there is no activity coefficient involved in the derivation. The main assumptions in Eq. 2.20 are:

- the Einstein relation is valid for a concentrated solution;
- the external electrical field is constant.

Compared with the assumptions in Fick's law (i.e. Eq. 2.8), the above assumptions seem more reasonable.

Similarly, for a solution–concrete system,

$$J_{0m} = D_0 \cdot \frac{zFE}{RT} \cdot c \tag{2.21}$$

2.5.2 Steady-state migration

In the case of a steady-state process, i.e. a constant flux, c is independent of distance and time, provided that the chloride concentration in the bulk solution remains constant. Therefore, Eq. 2.21 can be rewritten as:

$$D_0 = \frac{J_{0m} RT}{zFEc_0} \tag{2.22}$$

where E is the density of the electric field. Alternatively, if the area of concrete is used for calculation of the flux,

$$D_{ssm} = \frac{RT}{zFEc_0} \cdot J_{concrete} = \frac{RT}{zFEc_0} \cdot J_{0m} \cdot p_{sol} = D_0 \cdot p_{sol} = D_{ss} \tag{2.23}$$

It can be seen that the diffusion coefficient obtained from a steady-state migration test is identical to that obtained from a steady-state diffusion test.

It is worth pointing out that c_0 in Eq. 2.23 is the number of charged particles, i.e. the ionic concentration. In some papers (e.g. Andrade (1993)), it has been defined as activity, i.e. the product of the concentration and an activity coefficient that describes the thermodynamic behaviour of the ions.

However, in our opinion, the activity coefficient can influence only the diffusion process, and not the migration process, because under the action of a strong external electrical field, the ions' own thermodynamic behaviour becomes negligible and all charged particles will move following the specified directions of the electrical field.

2.5.3 Non-steady-state migration

For a non-steady-state process,

$$\frac{\partial c_t}{\partial t} = \frac{\partial c}{\partial t}\left(1 + \frac{\partial c_b}{\partial c}\right) = -\frac{\partial J_{0m}}{\partial x} = -D_0 \cdot \frac{zFE}{RT} \cdot \frac{\partial c}{\partial x} \tag{2.24}$$

or

$$\frac{\partial c}{\partial t} = -\frac{D_0}{1 + \dfrac{\partial c_b}{\partial c}} \cdot \frac{zFE}{RT} \cdot \frac{\partial c}{\partial x} = -D_{nssm} \cdot \frac{zFE}{RT} \cdot \frac{\partial c}{\partial x} \tag{2.25}$$

For a certain test duration t, the chloride front moves to a distance x_f, and c at the chloride front can reach a maximum of c_0, the concentration in the bulk solution. Thus Eq. 2.25 can be written as:

$$D_{nssm} = \frac{D_0}{\left(1 + \dfrac{\displaystyle\int_0^{c_b} dc_b}{\displaystyle\int_0^{\gamma_b c_0} dc}\right)} = \frac{RT}{zFE} \cdot \frac{\displaystyle\int_0^{x_f} dx}{\displaystyle\int_0^{t} dt}$$

resulting in

$$D_{nssm} = \frac{D_0}{\left(1 + \dfrac{c_b}{\gamma_b c_0}\right)} = \frac{RT}{zFE} \cdot \frac{x_f}{t} \tag{2.26}$$

where $0 \le \gamma_b \le 1$, which can be thought of as an effective coefficient for chloride binding during migration. Under certain experimental conditions, γ_b might be constant.

According to Eq. 2.24,

$$\frac{c_b}{\gamma_b c_0} = \frac{f_b}{1000} \cdot \frac{W_{gel}}{p_{sol}} \cdot \frac{(\gamma_b c_0)^\beta}{\gamma_b c_0} = \frac{f_b}{1000} \cdot (\gamma_b c_0)^{\beta-1} \cdot \frac{W_{gel}}{p_{sol}}$$

where f_b is the binding constant, β is a constant and W_{gel} is the weight of the gel, or

$$\frac{c_b}{\gamma_b c_0} = K_b \cdot \frac{W_{gel}}{p_{sol}} \tag{2.27}$$

where

$$K_b = \frac{f_b}{1000} \cdot (\gamma_b c_0)^{\beta-1} \tag{2.28}$$

Thus

$$D_{nssm} = \frac{D_0}{\left(1 + K_b \cdot \dfrac{W_{gel}}{p_{sol}}\right)} = \frac{RT}{zFE} \cdot \frac{x_f}{t} \qquad m^2/s \tag{2.29}$$

In contrast to D_{nss} in Eq. 2.15, the diffusion coefficient obtained from a non-steady-state migration test, D_{nssm}, is relatively constant.

2.6 Diffusion and migration

2.6.1 *Combined diffusion and migration*

In cases where diffusion and migration occur simultaneously, the flux should be the sum of the contributions of each mechanism. Thus, combining Eqs 2.12 and 2.21 for this case results in a flux of ions through a unit area of solution J_0 of

$$J_0 = J_{0d} + J_{0m} = -D_0 \frac{\partial c}{\partial x} + D_0 \cdot \frac{zFE}{RT} \cdot c$$

or

$$J_0 = -D_0 \left(\frac{\partial c}{\partial x} - \frac{zFE}{RT} \cdot c \right) \tag{2.30}$$

2.6.2 Steady-state process of diffusion and migration

For a steady-state process, the analytical solution to Eq. 2.30 has been given by Tang and Nilsson (1993a) as:

$$D_0 = \frac{J_0 \, RTL \left(e^{\frac{zFU}{RT}} - 1 \right)}{zFU \left(c_0 \, e^{\frac{zFU}{RT}} - c_1 \right)} \tag{2.31}$$

where c_0 and c_1 are the chloride concentrations in the cell supplying chloride ions (upstream cell) and in the cell collecting them (downstream cell), respectively. It is obvious that, if the electrical potential difference U is large enough to make $\exp(zFU/RT) \gg 1$, and $c_0 \geq c_1$, the above equation is reduced to Eq. 2.22. If U is very small, it is reduced to Eq. 2.12, with $-(\partial c/\partial c) = (c_0 - c_1)/L$.

2.6.3 Non-steady-state process of diffusion and migration

For a non-steady-state process,

$$\frac{\partial c_t}{\partial t} = \frac{\partial c}{\partial t}\left(1 + \frac{\partial c_b}{\partial c}\right) = -\frac{\partial J_0}{\partial x} = D_0 \left(\frac{\partial^2 c}{\partial x^2} - \frac{zFE}{RT} \cdot \frac{\partial c}{\partial x}\right) \tag{2.32}$$

or

$$\frac{\partial c}{\partial t} = \frac{D_0}{1 + \dfrac{\partial c_b}{\partial c}} \cdot \left(\frac{\partial^2 c}{\partial x^2} - \frac{zFE}{RT} \cdot \frac{\partial c}{\partial x}\right) \tag{2.33}$$

If the migration is dominant,

$$\frac{\partial c}{\partial t} = D_{nssm} \cdot \left(\frac{\partial^2 c}{\partial x^2} - \frac{zFE}{RT} \cdot \frac{\partial c}{\partial x}\right) \tag{2.34}$$

As shown in Eq. 2.29, D_{nssm} is relatively constant and has the same dimensions as an intrinsic diffusion coefficient D_0, i.e. m_x^2/s. In this case, the analytical solution to Eq. 2.34 can be derived, as reported by Tang and Nilsson (1992, 1993a) and Tang (1996a), as:

$$c = \frac{c_0}{2} \cdot \left[e^{ax} \cdot \mathrm{erfc}\left(\frac{x + aD_{nssm} \cdot t}{2\sqrt{D_{nssm} \cdot t}} \right) + \mathrm{erfc}\left(\frac{x - aD_{nssm} \cdot t}{2\sqrt{D_{nssm} \cdot t}} \right) \right] \qquad (2.35)$$

where a is a factor of the electrical potential $(a = zFE/RT)$ and erfc is the complement to the error function erf, $\mathrm{erfc} = (1 - \mathrm{erf})$. When the electrical field U/L is large enough and the penetration depth x_d is sufficient $(x_d > aDt)$, then

$$e^{ax_d} \cdot \mathrm{erfc}\left(\frac{x_d + aDt}{2\sqrt{Dt}} \right) \to 0$$

and the above equation becomes

$$c_d = \frac{c_0}{2} \mathrm{erfc}\left(\frac{x_d - aD_{nssm} \cdot t}{2\sqrt{D_{nssm} \cdot t}} \right) \qquad (2.36)$$

or

$$\frac{x_d - aD_{nssm} \cdot t}{2\sqrt{D_{nssm} \cdot t}} = \mathrm{erf}^{-1}\left(1 - \frac{2c_d}{c_0} \right) \qquad (2.37)$$

where c_d is the chloride concentration at which the colour changes when using a colorimetric method to measure x_d, and erf^{-1} is an inverse error function. Letting

$$\xi = \mathrm{erf}^{-1}\left(1 - \frac{2c_d}{c_0} \right)$$

Eq. 2.37 can be rewritten as

$$\left(\sqrt{D_{nssm}} \right)^2 + \frac{2\xi}{a\sqrt{t}}\sqrt{D_{nssm}} - \frac{x_d}{a \cdot t} = 0 \qquad (2.38)$$

Solving the above equation yields:

$$D_{nssm} = \frac{1}{a \cdot t}\left(\frac{2\xi^2}{a} + x_d - \frac{2\xi}{\sqrt{a}}\sqrt{\frac{\xi^2}{a} + x_d} \right) \qquad (2.39)$$

Under normal laboratory conditions $a >> \xi^2$. Equation (2.37) can be simplified as:

$$D_{nssm} = \frac{1}{a \cdot t}\left(x_d - \frac{2\xi}{\sqrt{a}}\sqrt{x_d} \right) \tag{2.40}$$

or

$$D_{nssm} = \frac{RT}{zFE} \cdot \frac{x_d - \alpha\sqrt{x_d}}{t} \tag{2.41}$$

where α can be taken as a laboratory constant,

$$\alpha = 2\sqrt{\frac{RT}{zFE}} \cdot \text{erf}^{-1}\left(1 - \frac{2c_d}{c_0} \right) \tag{2.42}$$

Comparing Eq. 2.41 with Eq. 2.29, one can see that the relationship between an average penetration front x_f and a penetration depth x_d is

$$x_f = x_d - \alpha\sqrt{x_d} \tag{2.43}$$

The latter can be measured by using a colorimetric method (see, for example, Collepardi *et al.* (1970), Otsuki *et al.* (1992) and Collepardi (1995)).

2.7 Other mechanisms

The above discussions about diffusion and migration are limited to a stationary liquid system. In other words, there is no water movement or exchange within the concrete. In reality, however, concrete structures may be exposed to different environments where a gradient of water pressure or vapour pressure may exist. In this case the following transport processes other than diffusion or migration may occur:

- hydraulic flow of chloride solution due to a gradient of water pressure (e.g. Edvardsen (1995));
- capillary suction of chloride solution in an unsaturated pore system due to the surface tension of pore walls (e.g. Collepardi and Biagini (1989), Volkwein (1991) and Akita and Fujiwara (1995));
- convection of chloride solution due to 'wick action' (e.g. Buenfeld *et al.* (1995));

- moisture flow and evaporation due to a gradient of vapour pressure (e.g. Tuutti (1982) and Ohama *et al.* (1995)).

The literature (e.g. Nilsson *et al.* (1996)) shows that it is difficult or even impossible to quantify the effects of these combined processes on the chloride transport. Numerous data should be collected and a better understanding is needed before this quantification becomes possible.

3   Test methods and their precision

3.1 Introduction

Owing to the important role of chloride ingress in concrete with regard to the durability of concrete structures, many methods have been proposed for testing such chloride ingress. However, as described in Chapter 2, the complexity of the process has, until now, hindered the reaching of a general agreement on a single test method for standardisation. Castellote and Andrade (2006) through an international network, RILEM TC 178-TMC Testing and Modelling Chloride Penetration in Concrete, collated details of various laboratory test methods, most of which were reviewed by Nanukuttan (2007). These methods can be categorised in different ways according to the test principles, e.g. diffusion or migration, steady-state or non-steady-state, conventional or accelerated, or laboratory or field applicable. Some commonly used test methods are reviewed in the following sections.

3.2 Conventional test methods

The term 'conventional test' generally refers to a method that simulates the natural process of chloride transport, such as diffusion cell tests and immersion tests, with a chloride concentration similar to that of seawater or de-icing salt. As diffusion is a very slow process, these conventional methods are very time consuming. It often takes months or years to obtain the test results.

3.2.1 Diffusion cell test

A diffusion cell test normally involves the following procedures:

- A thin slice of specimen is used to separate two cells, as shown in Figure 3.1, where the upstream cell contains chloride solution and the down-stream cell initially does not contain chloride ions.
- The chloride concentration in the solution of the down-stream cell is monitored at a certain interval of time.

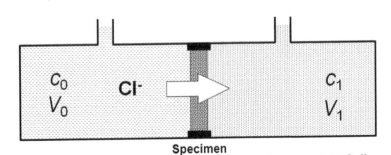

Up-stream Cell   **Specimen (with seal ring)**   **Down-stream Cell**

Figure 3.1 Experimental arrangement for a diffusion cell test.

- When a constant increase in the chloride concentration (i.e. a linear relationship between c_1 and t, implying a steady-state chloride flow) is observed, as in Figure 3.2, the experiment can be terminated and the diffusion coefficient can be calculated according to Fick's first law.

Usually an equation similar to Eq. 2.8:2 is used to calculate the diffusion coefficient under the steady-state, that is:

$$D_{ss} = J_{concrete} \cdot \frac{L}{c_0 - c_1} = \frac{V_{cell1}}{A_{concrete}} \cdot \frac{\Delta c_1}{\Delta t} \cdot \frac{L}{c_0 - c_1} \qquad \frac{m^3_{solution} \cdot m_x}{m^2_{concrete} \cdot s} \qquad (3.1)$$

where $\Delta c_1 / \Delta t$ is the average slope of the linear portion of the concentration–time curve. If $c_0 \gg c_1$, the latter is negligible. Alternatively, integrating Eq. 3.1 in the range from t_n to t yields:

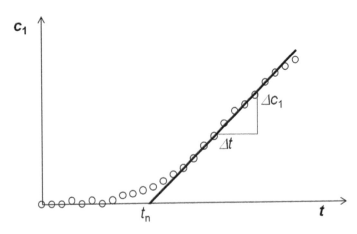

Figure 3.2 Example of a concentration–time curve obtained in a diffusion cell test.

$$D_{ss} = \frac{V_{cell1}}{A_{concrete}} \cdot \frac{L}{t - t_n} \cdot \left[\ln c_0 - \ln(c_0 - c_1) \right] \qquad \frac{m^3_{solution} \cdot m_x}{m^2_{concrete} \cdot s} \qquad (3.2)$$

where t_n is the intersection of the linear regression line (see Figure 3.2) on the t axis, and is called the 'time-lag'. Any pair of values t and c_1 on the linear regression line can be used in Eq. 3.2.

It should be noted that some researchers simply multiply the concentration c_0 in the above equation by the activity coefficient γ (Castellote *et al.*, 2001b). From the basic equation of diffusion (Eq. 2.7 in Chapter 2), it is clear that, if the effect of the activity coefficient is to be taken into account, it should be by multiplying by the term

$$\left(1 + \frac{c_0}{\gamma} \cdot \frac{\partial \gamma}{\partial c_0} \right)$$

but not by using a simple multiplication of $\gamma \cdot$ by c_0.

As the cross-sectional area of concrete includes both water-accessible pores and impermeable solid materials, the diffusion coefficient D_{ss} is sometimes called the 'effective diffusion coefficient'. The water-accessible pores of the concrete specimen should be taken into account in order to convert the effective diffusion coefficient into an 'intrinsic diffusion coefficient', as shown in Eq. 2.11.

According to the diffusion of mathematics (Crank, 1975), the diffusion coefficient under the non-steady-state, D_{nss}, can simply be calculated from the time lag, i.e.

$$D_{nss} = \frac{L^2}{6t_n} \qquad \frac{m^2_x}{s} \qquad (3.3)$$

Since Page *et al.* (1981) reported this method for determining the chloride diffusion coefficient of cement pastes, many other researchers have used either this or a similar test set-up to study the chloride diffusion in cement paste or mortar specimens (see, e.g., Hansson *et al.*, 1985; Roy *et al.*, 1986). While the basic principle of the test procedure has been the same, variations in sample thickness, concentration of the ion source solution and the method used to precondition the specimen before the test make the results difficult to compare between different procedures.

The diffusion cell method is considered the conventional way to determine the ion transport by diffusion, but long test durations are required to achieve the steady-state condition, because diffusion is a slow process. Therefore, the test is best suited to the measurement of the ion diffusion coefficient of hardened cement paste, where thin discs (typically of 3–4 mm thick) are tested. In the case of cement mortar discs having a thickness of 4–10 mm, typical tests may last for several months rather than weeks. It may even take years to obtain the steady-state condition for concrete, especially when testing concrete with a low water/binder ratio (e.g. < 0.5), which is often the case for structures exposed to chloride environments. The very long test duration means that hydroxides and alkalis may leach out from test specimens during the test. In order to limit this effect, a saturated solution of calcium hydroxide and a suitable concentration of potassium hydroxide are added to the upstream and downstream solutions in the diffusion cell, respectively, to correspond to what the binder may produce.

To date, no standard test procedure is available for this test method.

3.2.2 Immersion and ponding tests

Both the immersion and the ponding tests usually involve the following procedure:

- All surfaces of a specimen except one are sealed to prevent multi-directional penetration.
- After pre-saturation with water, the specimen is immersed in, or ponded with, a solution containing chloride ions (Figure 3.3).
- After a certain period of immersion or ponding, either the chloride content at discrete depths from the exposed surface is determined using dust samples collected by drilling, or the chloride profile is obtained by grinding the specimen successively from the exposed surface (Figure 3.4) and then analysing the chloride content of each powder sample.

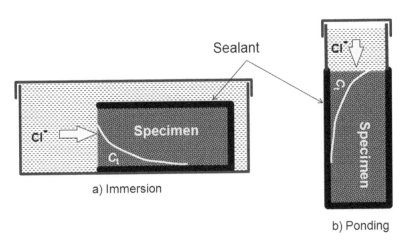

Figure 3.3 Experimental arrangement for the immersion test.

Figure 3.4 Example of grinding on a lathe (left) or a portable profile grinder (right)
to determine the chloride profile.

Under the assumption that there is no chloride binding, or that the bound
and free chloride ions follow a simple linear relationship with the line passing through the origin, as shown in Eq. 2.2, the diffusion coefficient can be
obtained by curve-fitting the chloride profile to the following equation, the
so-called 'error function solution', which is one of the solutions to Fick's
second law:

$$C_t = C_{ts} - (C_{ts} - C_{ti}) \cdot \mathrm{erf}\left(\frac{x}{2\sqrt{D_{nss} \cdot t}}\right)$$ (3.4)

where C_t is the content of total chloride by weight of material, C_{ti} is the
initial content of total chloride in concrete and C_{ts} is the content of total
chloride at the surface of the concrete, or, in the dimensionless form,

$$\frac{C_{ts} - C_t}{C_{ts} - C_{ti}} = \mathrm{erf}\left(\frac{x}{2\sqrt{D_{nss} \cdot t}}\right)$$ (3.4:1)

The chloride concentration in the exposure solution can close to that of
natural seawater (e.g. 3–6% NaCl, approximately 0.55–1.0 mol/l), or as high
as about 3 mol/l (165 g NaCl per litre), as used by Frederiksen (1992), to accelerate the diffusion process and shorten the test period. As will be discussed
in Chapters 4 and 5, due to the simplification of Fick's law when describing
chloride transport in concrete, the curve-fitted diffusion coefficient obtained
using the above equations is strongly dependent on the exposure period.

The immersion test has been widely used by many researchers since the 1970s (e.g. Collepardi *et al.* (1972)), although the curve-fitting technique was introduced later, with the development of personal computer. The earliest standardised ponding test, the AASHTO T259 procedure for determining the resistance of concrete to chloride ion penetration, was introduced by the American Association of State Highway and Transportation Officials (AASHTO), and has been in use since 1980. This was later standardised by the American Society for Testing and Materials (ASTM) as the ASTM C1543 procedure (since 1991). In this ponding test, a 3% NaCl solution is normally used to expose the test surface for a period of 90 days. The chloride sampling specified is, however, very rough, because the chloride content is determined from dust samples at large intervals (10 mm) from the exposed surface. In 2004, an immersion test was standardised in European countries as EN 13396. This method is basically similar to the AASHTO T259 (or ASTM C1543) test, except that the test specimens are immersed in the 3% NaCl solution and the exposure is done at an elevated temperature of 40°C. After exposing specimens for 28 days, 3 months and 6 months, they are sampled for the chloride content at three depths (0–2, 4–6 and 8–10 mm). Because of these large sampling intervals and the very few measurement points, it is difficult to calculate the diffusion coefficient with sufficient precision. Furthermore, the immersion periods of 90 days and 6 months are too long for it to be practicable to obtain the chloride penetration resistance of concretes for most construction projects.

In the mid-1990s, the Nordic countries standardised an accelerated immersion test as the Nordtest NT BUILD 443, which was published in 1995. In this standard method a concentrated chloride solution (165 g NaCl per litre) is used for an immersion period of 35 days. This Nordic procedure was also adapted by the ASTM as ASTM C1556 in 2004, in which the sampling intervals suggested by Tang and Sørensen (1998) were recommended. Currently, CEN (the European Committee for Standardisation) is standardising an immersion test EN/TS 12390–11, which is based on both NT BUILD 443 and ASTM C1556, with some modifications. The salient differences in the various immersion tests are listed in Tables 3.1 and 3.2.

3.3   Accelerated test methods

The term 'accelerated test' is often used to refer to those methods that accelerate the process of chloride transport by means of an external electrical field, a concentrated chloride solution, or an elevated temperature. As the acceleration due to concentrated chloride solutions (e.g. NT BUILD 443, as described in Section 3.2.2) or elevated temperatures is relatively limited, an external electrical field is often used in rapid test methods, which give a test duration of a few hours to less than a week. A number of rapid methods have been proposed since the beginning of 1980s, such as the rapid chloride permeability test (RCPT; more commonly known as the Coulomb test)

Table 3.1 The salient differences in the specimens and preconditioning used in different immersion tests

	ASTM C1543 (successor to AASHTO T 259)	NT BUILD 443	ASTM C1556	prEN/TS 12390-11
Specimen	Exposed surface area > 0.030 m², thickness 90 ± 15 mm	Dimension across the exposure surface > 70 mm, thickness >70 mm	Similar to NT BUILD 443	100 mm diameter cylinder or 100 mm cubes, thickness > 3 × maximum size of aggregate
Exposure surface	Unspecified, but understandably a finished surface	Cut surface after slicing off the topmost approximate 10 mm	Finished surface	Cut surface as in NT Build 443
Preconditioning	None (Chloride exposure direct after curing in moist condition until age 14 days, followed by storage at 23 ± 2°C and 50 ± 5% relative humidity until age 28 days)	Immersion in saturated calcium hydroxide water at 23 ± 2°C until the mass changes by < 0.1% in 24 hours	Similar to NT BUILD 443; alternatively, use of vacuum saturation similar to that described in ASTM C1202[1]	Vacuum saturation similar to that described in ASTM C1202, but under an absolute pressure of 10–50 mbar (1–5 kPa), and with distilled or demineralised water as the saturation liquid

Note

1 According to ASTM C1202-05, vacuum under an absolute pressure of <50 mmHg (6650 Pa or 66.5 mbar) for 3 hours, then with the vacuum pump still running, fill the vacuum chamber with cool boiled water to immerse the specimens; maintain the vacuum for a further hour before allowing air to re-enter the chamber, and keep the specimens in the water for further 18 ± 2 hours.

Table 3.2 The salient differences in exposure conditions, duration and sampling used in different immersion tests

	ASTM C1543 (successor to AASHTO T 259)	EN 13396	NT BUILD 443	ASTM C1556	prEN/TS 12390-11
Exposure conditions	Ponding in 3% (by mass) NaCl solution with a depth of 15 ± 5 mm, stored at 23 ± 2°C and 50 ± 5% relative humidity	Immersion in a solution containing 3% (by mass) NaCl at 40°C, stored at 23 ± 2°C and 50 ± 5% relative humidity	Immersion in a solution containing 165 g NaCl per litre, at 23 ± 2°C	Similar to NT BUILD 443	Immersion in a solution containing 3% (by mass) NaCl as reference, at 20 ± 2°C
Exposure duration	3 months, and subsequently after 6 and 12 months of ponding, and at 12-month intervals thereafter	3 months, and subsequently after 28 days, 3 months and 6 months of immersion	35 days	35 days	90 days as reference
Sampling	At intervals of 10–20, 25–35, 40–50 and 55–65 mm	At 0–2, 4–6 and 8–10 mm	At least 8 layers, with the first layer ≥ 1 mm; or a minimum of 6 layers covering the profile to the depth, if the chloride level is 0.03 mass% higher than the initial (background) value	Following recommended depth intervals (see Table A1 in the Appendix); otherwise similar to NT BUILD 443	Similar to the recommended depth intervals (see Table A1 in Appendix), but further layers should be ground off if the chloride level in the deepest layer is 0.015 mass% higher than the initial value

(Whiting, 1981), the potential diffusion (PD) index test (Dhir *et al.*, 1990) and the rapid migration test (Tang and Nilsson, 1992).

3.3.1 Rapid chloride permeability test (Coulomb test)

The RCPT (Coulomb test) was proposed by Whiting (1981). Owing to its simplicity and short testing period, this method has been adopted as a standard test (e.g. AASHTO T277 since 1983, and ASTM C1202 since 1994). This method involves the following procedure:

- A specimen of 50 mm thickness is placed between two cells similar to those in the diffusion cell test, but with an electrode in each cell, as shown in Figure 3.5.
- A potential of $U = 60\,\mathrm{V}$ is applied across the specimen, with the cathode in the cell containing chloride ions.
- The electrical current is monitored during the test.

The test results are expressed as the Coulomb value, i.e. the total charge which passes through the specimen during the first 6 hours of the test, which can be calculated from the measured current data using the following equation:

$$Q = 900\left(I_0 + 2I_{30} + 2I_{60} + \ldots + 2I_{300} + 2I_{360}\right) \tag{3.5}$$

where Q is the charge passed (coulomb), I_0 is the current (ampere) immediately after voltage is applied, and I_t is the current (ampere) at t min after voltage is applied.

From the Coulomb value the qualitative chloride ion penetrability can be estimated using the criteria listed in Table 3.3.

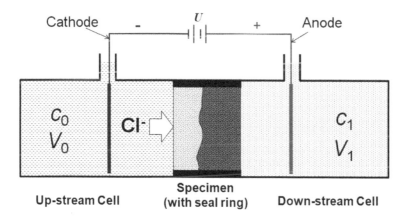

Figure 3.5 Experimental arrangement for a migration test.

Table 3.3 Chloride ion penetrability based on the charge passed (according to ASTM C1202)

Charge passed (coulomb)	Chloride ion penetrability
> 4000	High
2000–4000	Moderate
1000–2000	Low
100–1000	Very low
< 100	Negligible

This method cannot, however, give information about chloride diffusion in concrete, for the following reasons:

- The conductivity of the pore solution is influenced by the concentration of ions in the pore solution, and the concentration of these ions can vary between cement types (Page and Vennesland, 1983). Charge passed, however, is related to all ions transported through concrete, not just chloride ions. Therefore, the test accounts for the total current and not that corresponding to the chloride flux.
- The test only provides an index, and no theoretical relationship exists whereby the charge passed or current values obtained could be used to determine the chloride diffusivity (Stanish *et al.*, 2000). In many cases, the chloride ions do not penetrate through the entire thickness of the specimen during the 6-hour duration of the test.
- The temperature rise due to the high voltage (60 V) will significantly affect the conductivity of ions and, as a consequence, the total charge passed during the test.

As the hydroxyl ions have a significantly high conductivity, i.e. 5.26×10^{-9} m²/s in the bulk dilute solution, compared with that of all other ions, including the chloride ions, which have a conductivity of 2.03×10^{-9} m²/s in the bulk dilute solution, the current should mainly be contributed by the hydroxyl ions in the pore solution in concrete. As reported by Feldman *et al.* (1994), a similar Coulomb value can also be obtained by using solutions other than a chloride solution. Therefore, the Coulomb test is just a test for the conductivity of the pore solution, rather than a test for chloride transport properties.

3.3.2 Potential diffusion index test

The PD index is in fact a steady-state migration test. Migration cells similar to those used in the Coulomb test (see Figure 3.5) can be used in this test. To avoid the Joule effect, the potential is often lowered down to $U = 10$–12 V. In contrast to the Coulomb test, in the PD index test the chloride ions must be driven through the specimen, and the flow rate of the chloride ions is monitored in a manner similar to the diffusion cell test.

Dhir *et al.* (1990) tried to calculate the diffusion coefficient from this test by simply using Fick's first law (i.e. Eq. 3.1 or Eq. 3.2). Apparently, this has little theoretical support, and hence the coefficient obtained in this way is called 'potential diffusion index', or PD index.

The PD index test was adopted as a standard method (NT BUILD 355) in 1989. With the development of migration theory, this standard was revised in 1995 (Gautefall *et al.*, 1995).

3.3.3 Steady-state migration test

The set-up for a steady-state migration test is the same as that used in the PD index test described in Section 3.3.2. In 1993, Andrade (1993) and Tang and Nilsson (1993a) published similar equations for calculating the steady-state diffusion coefficient from a steady-state migration test. The difference in the equation between these two research groups is that Andrade (1993) used the chloride activity instead of the concentration in the migration equation. According to Eq. 2.23, the equation for calculating the steady-state diffusion coefficient from this test should be as follows:

$$D_{ssm} = \frac{RTL}{zFUc_0} \cdot J_{concrete} = \frac{RTL}{zFUc_0} \cdot \frac{V_1 \cdot \Delta c_1}{A_{concrete} \cdot \Delta t} \qquad \frac{m^3_{solution} \cdot m_x}{m^2_{concrete} \cdot s} \qquad (3.6)$$

As pointed out in Chapter 2, c_0 in the above equation is the number of charged particles, i.e. the concentration of the chloride solution, not the activity of the chloride ions. Therefore, according to Tang and Nilsson (1993a), no activity coefficient is needed in the calculation.

The non-steady-state diffusion coefficient can also be derived from this test (Castellote *et al.* (2001b)):

$$D_{nssm} = \frac{1}{t_{nm}} \cdot \left(\frac{L}{v}\right)^2 \cdot \left(v \coth\frac{v}{2} - 2\right) \qquad \frac{m_x^2}{s} \qquad (3.7)$$

where t_{nm} is the time lag from the steady-state migration test and

$$v = \frac{zFU}{RT} \qquad (3.8)$$

The first standard procedure for the steady-state migration test is the Nordic test method NT BUILD 355 (revised in 1995–97). In this standard method, an external potential of 12 V is applied across a concrete specimen of size Ø100 × 50 mm. In the method proposed by Castellote *et al.* (2001b), hereafter called the IETcc test, a specimen of size Ø75 × 20 mm or Ø100 × 20 mm is used. The chloride concentration c_1 in the downstream cell is determined at a certain interval by chemical titration in the NT BUILD 355 test, or by means of an indirect method (e.g. measurement of the conductivity, which is then converted to the concentration of chloride ions with the help of a calibration curve, as in the IETcc test).

3.3.4 Non-steady-state migration test

In early 1990s, Tang and Nilsson (1991, 1992, 1993a) proposed a rapid test based on non-steady-state chloride migration, called the CTH rapid test (CTH – Swedish abbreviation of Chalmers University of Technology) in the Scandinavian countries, and later known as the RCM (rapid chloride migration) test throughout the world.

   In this method, an external electrical field is applied to accelerate the chloride transport process of migration. The method involves applying a DC potential of 10–60 V, depending on the quality of concrete, across a specimen of size Ø100×50 mm for a duration of, in most cases, 24 hours, and afterwards measuring the chloride penetration depth using a colorimetric technique (e.g. Collepardi *et al.* (1970) and Otsuki *et al.* (1992)). The migration coefficient can be calculated from the measured chloride penetration depth, and this coefficient may further be used in durability designs. In normal cases, cut surfaces of the specimens are exposed to the solutions during the test.

   The method involves the following procedure:

- A specimen of size Ø100×50 mm is placed in a rubber sleeve, which forms a cell for holding the electrolyte (normally 0.3 M NaOH) and the anode (normally stainless steel plate).
- The clip-tightened rubber sleeve containing the specimen is placed above the cathode (normally a stainless steel plate) in a migration box containing chloride solution (normally 10% NaCl in saturated limewater), as shown in Figures 3.6 and 3.7. The specimen is titled to make it easy for gas bubbles at the cathode plate to escape.
- An external potential is applied across the specimen for a specified period (in most cases 24 hours).
- After the migration test, the specimen is split into two, and a silver nitrate (0.1 N) solution is sprayed onto the newly split surfaces. The average penetration depth is determined from measurements made at five different locations.
- Finally, the chloride migration coefficient is calculated from the measured average penetration depth, using the following equations:

$$D_{nssm} = \frac{RTL}{zFU} \cdot \frac{x_d - \alpha\sqrt{x_d}}{t} \tag{3.9}$$

where α can be taken as a laboratory constant,

$$\alpha = 2\sqrt{\frac{RTL}{zFU}} \cdot erf^{-1}\left(1 - \frac{2c_d}{c_0}\right) \tag{3.10}$$

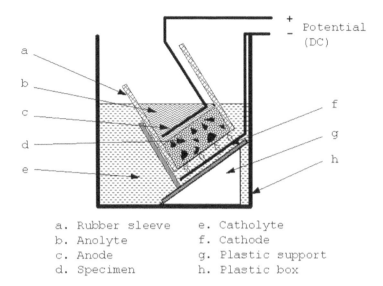

a. Rubber sleeve   e. Catholyte
b. Anolyte   f. Cathode
c. Anode   g. Plastic support
d. Specimen   h. Plastic box

Figure 3.6 Experimental arrangement for the RCM test.

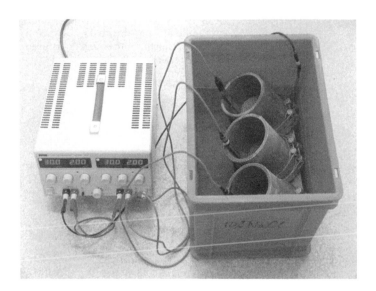

Figure 3.7 Example of test equipment used in the RCM test.

As there is a potential drop at the interface between the electrode and the electrolyte (McGrath and Hooton, 1996), the actual potential U applied across the specimen should be measured using two reference electrodes kept close to the two sides of the specimen. According to the data reported by McGrath and Hooton (1996), the sum of the drop in potential at the two electrodes is about 2 V. To simplify the measurement, a total drop of 2 V can be subtracted from the applied potential measured between the two electrodes, i.e. $U = U_0 - 2$, where U_0 is the applied potential measured between the two electrodes.

Owing to its simplicity and short test period, the method has been used worldwide, and was adopted as the Nordic standard test NT BUILD 492 in 1999, the American provisional standard AASHTO TP 64 in 2003, a standard test in Swiss standard SIA 262/1-B in 2004, and in Chinese National Standard GB/T 50082 in 2009. Some minor differences between the various versions of the RCM test are listed in Table 3.4.

Table 3.4 Salient differences in various RCM tests

	NT BUILD 492	SIA 262/1	AASHTO TP64	CHLORTEST recommendation
Preconditioning	Vacuum saturation similar to that described in ASTM C1202, but under an absolute pressure of 10–50 mbar (1–5 kPa), and with saturated lime water as the saturation liquid	Immersion in water for 7 days (first up to the middle of the specimen for 24 hours and then completely submersed for 144 hours)	Similar to NT BUILD 492, but with de-aired tap water as the saturation liquid	Similar to NT BUILD 492, but with distilled or demineralised water as the saturation liquid
Exposure conditions	10% (by mass) NaCl as catholyte and 0.3 N NaOH as anolyte	3% (by mass) NaCl in 0.2 N KOH as catholyte and 0.2 N KOH as anolyte	Similar to NT BUILD 492	Similar to NT BUILD 492
Test conditions	Applied potential between two electrodes of $U_0 = 10$–60 V at 20–25°C for 24 hours or longer	Applied potential between two electrodes of $U_0 = 20$–30 V at room temperature for 24 or 16 hours	Applied potential between two electrodes of $U_0 = 10$–60 V at 23 ± 2°C for 18 hours	Similar to NT BUILD 492

3.3.5 Resistivity test

The resistivity test has sometimes been used as a measure of the transport property of concrete (McCarter *et al.*, 1995; Frederiksen *et al.*, 1997b; Andrade *et al.*, 2000a; Polder, 2001; Gjørv, 2009). This is a simple and quick test, and it can be applied to different sizes of specimens. A typical test arrangement is shown in Figure 3.8. The test often involves the following procedure:

- A specimen of known thickness, saturated with distilled water, artificial pore solution, or very concentrated NaCl solution, is placed between two conductive plates, using moist sponges as contact media.
- A constant alternating current at 1 kHz is applied between two electrodes and the potential between the electrodes is measured.
- A reference measurement is also made, i.e. making the same measurement under the same conditions but without a specimen.
- The resistivity of concrete is calculated using the following equation:

$$\rho = \frac{A}{L} \cdot \frac{U_{s+sp} - U_{sp}}{I} \qquad (3.11)$$

where ρ is the resistivity (Ωm) of concrete, A is the cross-sectional area (m^2) of the specimen, L is the thickness (m) of the specimen, I is the applied current (A), and U_{s+sp} and U_{sp} are the potential (V) between two electrodes measured with and without the specimen, respectively.

It is known that resistivity is inversely proportional to conductivity. The conductivity of concrete is dependent on both the water content of the con-

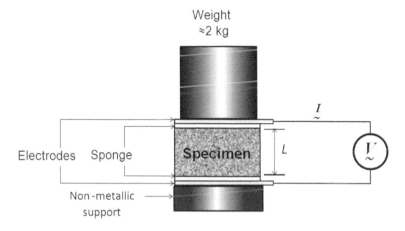

Figure 3.8 Example of a test arrangement for the resistivity test.

crete, or the porosity if the concrete is saturated, and on the types of ion in and the ionic concentration of the pore solution. The molar conductivity of the pore solution can be expressed as:

$$\Lambda_m = F\sum_i z_i c_i u_i \tag{3.12}$$

where Λ_m is the molar conductivity ($\Omega^{-1}\,m^2\,mol^{-1}$), and z_i, c_i and u_i are the valence, molar concentration and mobility of the ith ion, respectively. According to Nernst–Einstein equation,

$$D_i = \frac{RT}{z_i F^2}\cdot\lambda_i = \frac{RT}{z_i F^2}\cdot t_i\Lambda_m \tag{3.13}$$

where D_i, λ_i and t_i are the diffusion coefficient, molar conductivity and transference number of the ith ion, respectively. For porous materials such as concrete, the effective diffusion coefficient of the ith ion may be expressed as:

$$D_{effi} = \frac{RT}{z_i F^2}\cdot t_i\sigma = \frac{RT}{z_i F^2}\cdot\frac{t_i}{\rho} \tag{3.14}$$

where σ is the conductivity of concrete. If the pore solution contains chloride ions at a concentration of c_{Cl}, its transference number will be:

$$t_{Cl} = \frac{z_{Cl}c_{Cl}u_{Cl}}{\sum_i z_i c_i u_i} = \frac{\lambda_{Cl}}{\Lambda_m} = \frac{I_{Cl}}{I} \tag{3.15}$$

where t_{Cl} is the transference number of the chloride ions in the pore solution, I is the current applied in the resistivity or conductivity test, and I_{Cl} is the current contributed by the chloride ions. The value of I_{Cl} cannot be measured exactly, but can be estimated if the compositions of the pore solution in the concrete specimen is homogeneous and the concentrations of chloride and all other ions are known. In most cases, however, specimens used for the resistivity test contain no chlorides, or the concentrations of chloride and other ions are unknown, or the chloride ions are not homogeneously distributed in the specimen. In the latter case, the measured resistivity may not represent the average ion distribution but represent instead a portion where there are fewer or no chloride ions. Therefore, due to the unknown chloride transference number, it is theoretically weak to relate the chloride diffusion coefficient to the measured resistivity.

As the resistivity test is very simple and quick to perform, it is always possible to establish empirical relationships between the results of the resistivity test and various diffusion or migration tests. Therefore, this test is very suitable for quality control in the production of concrete with known mixture proportions and raw materials.

3.4 Test methods for *in situ* applications

For new structures, the ingress of chloride ions into concrete can be monitored using sensors embedded during construction. For existing concrete structures, samples can be taken from different depths for analysis of the chloride content. There are also test methods that can be applied equally to new as well as existing concrete structures. The selection of a test method will depend on the objective of the assessment. If the objective is to assess the condition of the structure, a single-point measurement of the chloride content at a certain depth from the exposed surface and the electrical resistance of the cover concrete can be used. However, if the intention is to use the results in predictive models, then tests that provide transport parameters need to be used, such as a multi-point measurement of chlorides to establish the chloride profile and *in situ* migration tests. Some of these *in situ* methods are described in this section.

3.4.1 Use of chloride concentration profiles in concrete

In order to determine the distribution of the chloride ion concentration or content in concrete, powder samples are collected from various depths from the exposed surface, and perpendicular to it, and analysed using standard procedures. The chloride ion concentrations or contents thus obtained are expressed conventionally as the percentage of chloride ions by mass of concrete sample or binder, if the content of the latter is known. A chloride profile is obtained by plotting the chloride concentration or content data against the corresponding depth from where the samples were collected. The method of analysis is similar to that used in NT BUILD 443, and fitting a non-linear regression curve for Eq. 3.4 enables the determination of the non-steady-state diffusion coefficient.

The chloride ions have to penetrate into the concrete in order to establish a chloride profile. Therefore, this approach cannot give any value for new concretes. In other words, until there is a build-up of chloride concentration or content inside the concrete, the chloride diffusion coefficient cannot be determined for new constructions. For concrete structures exposed to either non-severe conditions or to severe conditions but for a very short duration, errors associated with sampling and curve fitting could also lead to misinterpretation of the results (Nanukuttan *et al.*, 2004). The reason for this is that, in these cases, chloride profiles are rarely developed, and most data points would show zero chloride concentration or content.

As in the case of immersion tests described in Section 3.2.2, the degree of saturation of the concrete in the structure, the concentration of the salt solution to which structures are exposed and the temperature of the exposure environment influence the rate of chloride ingress. Therefore, any generalisation of the non-steady-state diffusion coefficient from *in situ* chloride profiles is inappropriate, and values should be reported along with the corresponding environmental factors.

Single-point measurement of chloride concentration or content in concrete

The chloride concentration or content at the level of reinforcement can provide an indication of the extent of damage already caused to the structure. This type of single-point measurement can be used to compare the performance of different concrete mixes. If the sample collected for chloride analysis represents more aggregate than cement paste, and if the cement content in the sample was not analysed, this would lead to an incorrect interpretation of the results. Furthermore, this method cannot provide any information on the rate of chloride ingress into the concrete, and hence the data cannot be used in any service-life prediction models.

Multi-point measurement of chloride concentration
or content in concrete and non-linear regression analysis

Multi-point measurement of the chloride concentration or content in concrete is required to obtain a chloride profile. The depths at which concrete powder is collected depend on the composition of the concrete in the structure. It is important to ensure that at least six single-point measurements are made between the concrete surface and the point at which the chloride concentration or content is negligible, as specified in NT BUILD 443. The first few millimetres of concrete can be affected by a convective flow, as shown in Figure 3.9. Therefore, it is necessary to avoid data from this layer

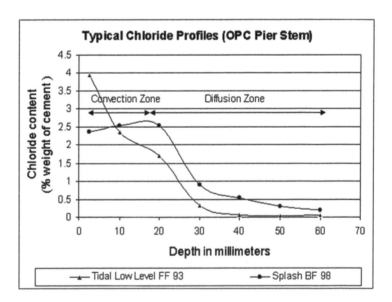

Figure 3.9 A typical chloride profile from OPC concrete piers exposed to the tidal and splash Zone (Nanukuttan *et al.*, 2008). The diffusion and convection zones for the pier exposed to the wetting and drying conditions in the splash zone can be seen.

for curve fitting purposes, in order to obtain a more realistic diffusion coefficient (Nanukuttan *et al.*, 2008).

Non-linear curve fitting of this profile, through Fick's second law (Eq. 3.4), will yield a diffusion coefficient and surface chloride concentration. These two parameters are conventionally used in service-life prediction models, and thus accuracy of the curve fitting is of vital importance. A linear approximation method for curve fitting (Poulsen, 1990) is also available, but it is generally considered less accurate. The error function solution as outlined in NT BUILD 443 (see Eq. 3.4) can be used for non-linear curve fitting with a greater accuracy.

3.4.2 Use of accelerated chloride migration tests

As described in Sections 2.5, 2.6 and 3.3, an external electric field can greatly accelerate the transport of chloride ions. Therefore, some of the accelerated tests described in Section 3.3 can also be performed on concrete *in situ* without extracting cores from structures for laboratory testing.

In situ *rapid chloride permeability test (Coulomb test)*

Whiting (1981) developed an *in situ* rapid chloride permeability test based on the same principle as the laboratory-based rapid chloride permeability test (AASHTO T277 or, later, ASTM C1202). A schematic drawing of the apparatus developed by Whiting (1981) is shown in Figure 3.10. Although the test is no longer used due to various limitations described below, it is briefly introduced here for completeness of the discussion.

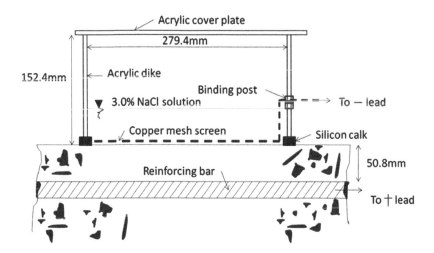

Figure 3.10 In situ rapid chloride permeability test. (Based on Whiting (1981).)

In this test, an acrylic dyke is attached to the surface of the concrete with a silicone caulking compound, and the silicone left to set overnight. Prior to testing, the concrete is preconditioned by vacuum saturating with limewater at 60°C for a period of 18 hours. At the start of the test, a 3% NaCl solution is poured into the dyke. A copper mesh screen is used as the cathode and a DC power supply applies a voltage between the mesh and the reinforcement. The test is carried out at 80 V for a period of 6 hours. The charge passed is used as an index to characterise the chloride diffusivity of the concrete, as in the laboratory-based test described in Section 3.3.1.

The disadvantages of the laboratory version discussed in Section 3.3.1 are also applicable to the *in situ* method. The temperature rise due to the high voltage (80 V in the *in situ* test rather than 60 V used in the laboratory test) also significantly affects the charge passed during the test, making the values obtained from the *in situ* test of no practical relevance. In addition, the thickness of the concrete cover influences the test results; for example, if a cover of 50 mm is assumed, results may vary by as much as 25% if the cover actually varies by 25 mm from the assumed value (Whiting, 1981).

In situ *rapid chloride migration test*

Similar to the *in situ* rapid chloride permeability test proposed by Whiting (1981), the RCM test described in Section 3.3.4 can also be used *in situ* to test the in-field performance of concrete, e.g. the reinforcement spacers (Tang, 2006). The background to the use of the RCM test *in situ* lies in the quality requirement for the reinforcement spacers used in an infrastructure, whereby the spacer and its bonding interface with the bulk concrete should have a chloride resistance similar to or better than that of the bulk concrete. It was found that coring test specimens from the structure introduced damage to the interface between the spacer and the concrete. Therefore, the RCM test on the *in situ* concrete containing reinforcement spacers was considered in order to examine the effectiveness of the bond between the spacer and the concrete.

A migration cell, as used in AASHTO T277, is placed at the position where a spacer lies (Figure 3.11). The method can be applied to both vertical and horizontal (top side) surfaces of a concrete structure, as shown in Figures 3.12 and 3.13. After tightening the cell on the concrete surface as shown in Figures 3.12 and 3.13, it is filled with 10% NaCl solution. An external potential of 20–80 V, or even higher, depending on the quality of the concrete, the cover thickness and what is required to achieve an acceptable test duration, is applied between the cell electrode (cathode) and the reinforcement (anode) in the concrete. After the migration test, a core is taken from the test position. The core can be axially sawn, either *in situ* or in the laboratory, in order to obtain better surfaces for the examination of the chloride penetration front by means of the colorimetric technique (as described in Section 3.3.4). The quality of the spacer and its bonding with the concrete

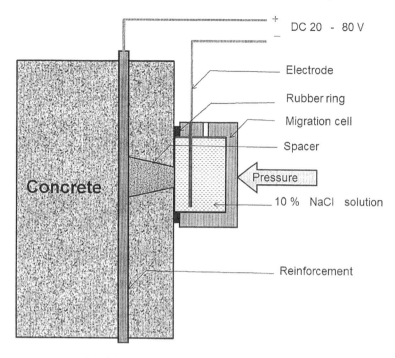

+ −
DC 20 - 80 V

Electrode

Rubber ring

Migration cell

Spacer

Pressure

10 % NaCl solution

Concrete

Reinforcement

Figure 3.11 Principle of the *in situ* RCM test (Tang, 2006).

Figure 3.12 Example of the *in situ* RCM test on a vertical concrete surface.

Figure 3.13 Example of the *in situ* RCM test on a horizontal (top side) concrete
surface.

can be immediately seen from the chloride penetration front (Figure 3.14).
The migration coefficient of the concrete can be estimated using Eq. 3.9, but
only approximately, due to the lack of preconditioning.

The disadvantage of this method is that a core must be taken from the test
position and, consequently, the core hole needs to be filled with repair concrete. However, this is not a big problem for most structures, where minor
damage to the facade is not an issue.

Permit ion migration test

This test has been developed based on the principle of the steady-state
migration of chloride ions through the concrete cover (Basheer *et al.*, 2005;
Nanukuttan *et al.*, 2009). Unlike conventional steady-state migration tests,
both the anolyte and the catholyte are placed on the concrete surface in
the form of concentric cylindrical reservoirs (Figure 3.15). This allows the
test to be performed on the concrete surface and thus avoids the need to
remove a specimen of concrete from the structure. The chloride ions from
the catholyte move towards the anolyte through the concrete cover due to
the application of an electric field. The amount of chloride ions arriving at
the anolyte (initially de-ionised water) is monitored regularly by means of
the change in conductivity of the solution. Although there is provision for
users to remove anolyte to assess the chloride concentration, the conductivity measurements allow the automation of test procedure. The conductivity measurements are then converted to equivalent chloride concentrations

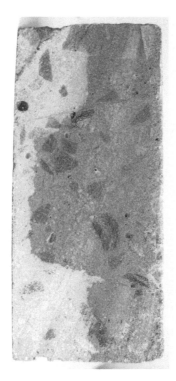

Figure 3.14 Example of chloride penetration in concrete with a spacer after applying the *in situ* RCM test. (Left) good bonding between the spacer and the concrete; (right) poor bonding.

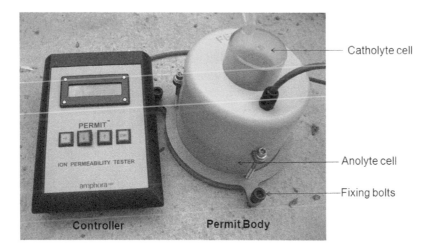

Figure 3.15 The *in situ* Permit ion migration test (Nanukuttan *et al.*, 2009).

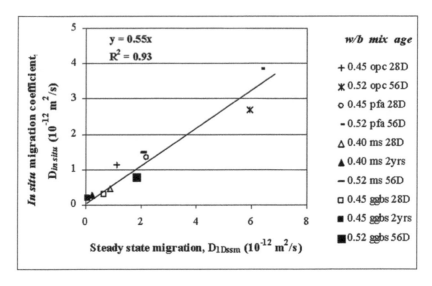

Figure 3.16 Relationship between the *in situ* migration coefficient and the steady-state migration coefficient (Nanukuttan *et al.*, 2009). GGBS, ground granulated blast furnace slag; MS, microsilica; OPC, ordinary Portland cement, PFA, pulverised fuel ash or fly ash, w/b, water/binder ratio.

using a calibration equation, and a concentration–time graph is plotted. The straight-line portion of the concentration–time graph is used to calculate an *in situ* migration coefficient, using a modified Nernst–Planck equation (Nanukuttan *et al.*, 2009).

A good linear correlation exists between the *in situ* migration coefficient obtained from the Permit ion migration test and the steady-state diffusion coefficient (Nanukuttan *et al.*, 2009), the steady-state migration coefficient, and the non-steady-state migration coefficient from laboratory-based migration tests (Figures 3.16 and 3.17). The peak current measured in the Permit ion migration test also correlates well with the bulk electrical resistivity of the concrete specimens (Nanukuttan *et al.*, 2007), and a modified Nernst–Einstein equation can be used to calculate the migration coefficient from the peak current (Nanukuttan *et al.*, 2009).

However, the limitation of the test is that the use of NaCl as the catholyte solution results in the test area becoming saturated with chloride ions. Although the ions can be removed by reversing the polarity of the electrodes, this is a time-consuming process and adds to the cost of carrying out the test. Therefore, non-corrosive ions need to be used as a substitute for chloride ions in order to make this test entirely risk free for the construction industry. The average test duration is 4–10 hours for normal concretes with water/binder ratios between 0.4 and 0.6, and 10–24 hours for concretes with water/binder ratios lower than 0.4. A completely automated version of this

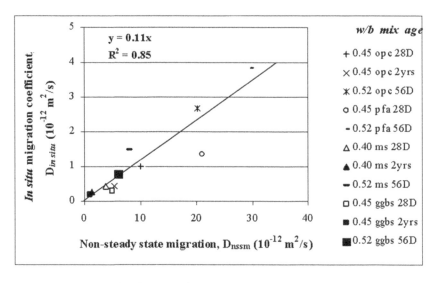

Figure 3.17 Relationship between the *in situ* migration coefficient and the NT Build 492 (Nordtest, 1999) non-steady-state migration coefficient (Nanuku-ttan *et al.*, 2007). GGBS, ground granulated blast furnace slag; MS, microsilica; OPC, ordinary Portland cement, PFA, pulverised fuel ash or fly ash; w/b, water/binder ratio.

test is commercially available, and the test can be performed using any of the three different test voltages to match the type of concrete used in a structure, and hence a wide range of concrete mixes can be tested.

3.4.3 Use of electrical resistance measurements

Concrete is a porous material, often with a well-developed pore network, and the pores are filled with highly conductive pore fluid. The ability of harmful ions to penetrate through this pore system may be characterised by the ability of the pore system to conduct electricity.

Embedded electrodes

McCarter *et al.* (1995) developed an electrode array system composed of miniature multi-electrodes to monitor the advance of a chloride front in concrete. A typical array system is shown in Figure 3.18.

Each probe consists of 10 stainless steel electrode pairs, 1.5 mm in diameter, mounted on a Perspex plate. The 5 mm tips of the electrodes are exposed, while the remainder is sleeved. All the electrodes are kept 5 mm apart in both vertical and horizontal directions, and each pair is staggered to minimise the interference caused by the aggregates on the resistivity measurements. The electrode

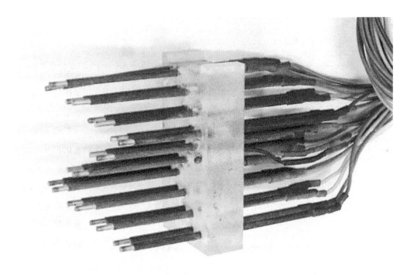

Figure 3.18 A Covercrete electrode. (From McCarter *et al.*, (1995).)

array is placed in the cover zone while the concrete is being cast in a structure.

The electrode system is able to provide information on the short- and long-term depth-related variations in the resistivity of concrete subjected to different environments. McCarter *et al.* (1996) showed that the resistivity measurements from embedded electrodes can used to monitor the ingress of water and chlorides into concrete; i.e. instead of using the resistivity measurements to determine the diffusivity, they suggested using the changes in resistivity as a monitoring technique. As the electrode pairs are placed at varying depths from the exposed face of a concrete structure, the resistivity of the concrete at different depths can be determined, which allows the advance of a chloride front to be monitored.

The main limitation of the electrode array system is that it is not possible to identify whether the change in resistivity is due to the ingress of a water front or a chloride front. Furthermore, chemical changes in concrete can also change the resistivity. This technique is not applicable to existing structures, as the electrodes need to be embedded during casting. Other similar electrode arrays have been introduced by different researchers. Although these can be used to monitor changes in resistivity, their primary application has been in monitoring the corrosion of steel in concrete.

Wenner four-probe resistivity test

The Wenner four-probe resistivity test determines the near-surface resistivity of a material using a four-probe array of electrodes (Figure 3.19) (Millard

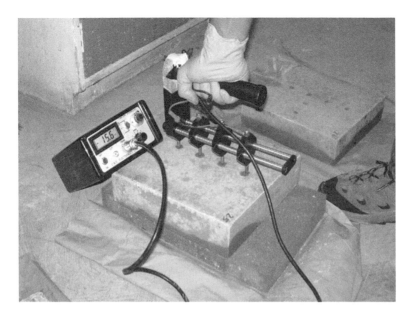

Figure 3.19 Wenner four-probe resistivity test.

et al., 1990). The technique consists of passing a constant magnitude alternating current between the two outer electrodes and measuring the potential difference across the two inner electrodes. The resistivity of the material is given by:

$$\rho_{\text{Wenner}} = 2\pi a \left(\frac{v}{I} \right) \tag{3.16}$$

where ρ_{Wenner} is the Wenner probe surface resistivity ($\Omega \cdot m$ or $k\Omega cm$), a is the spacing of the probes (m or cm), v is the potential difference measured across the electrodes (V or mV), and I is the applied current (A).

However, the test method has the following disadvantages:

- The contact between the electrodes and the surface is often the main source of error. To ensure the flow of current from the electrode to the concrete, a highly conductive gel is sometimes applied to the tip of the electrode. If this gel is not applied correctly, an inaccurate result may be obtained. An alternative is to use wetted wooden plugs inserted into the electrodes to ensure good contact between the electrodes and the concrete surface.
- In order to determine the diffusivity of chloride ions in concrete using the resistivity, the concentrations of the different ions in the pore solution should be known, as discussed in Section 3.3.5. It may, however,

be possible to make a rough estimate of the diffusivity of chloride ions in concrete by using the resistivity values (Polder, 2001).

3.4.4 Discussion of in situ methods

Some of the *in situ* test methods that can provide information about either the condition of the structure or the transport parameter (e.g. chloride diffusion coefficient), which can be used to predict further ingress of chloride ions into concrete, have been described in Sections 3.4.1 to 3.4.3. The single-point measurement of chloride concentration was seen to be useful for assessing the condition of a structure or the influence of the chloride concentration on the condition of the reinforcement. If more information is required regarding the rate of ingress of chloride ions, a full chloride profile will have to be obtained using multi-point measurement of the chloride concentrations. In general, both single- and multi-point measurements can only be performed in a structure that has been exposed to the chloride environment for a certain number of years.

Accelerated chloride migration tests can be performed on both new and existing structures. Several limitations of the *in situ* rapid chloride permeability test were identified, including the difficulty of using the charge passed as a measure of the chloride ion transport coefficient. The *in situ* RCM test is a useful tool for examining the quality of concrete, including the quality of reinforcement spacers, although this method is partly destructive due to the need for coring. The *in situ* migration coefficient determined using the Permit ion migration test has been proven to correlate well with both diffusion and migration coefficients obtained by means of various standard laboratory-based test methods. The only limitation of the Permit ion migration test is that, as the test uses chloride ions, the test area will become contaminated with chloride ions.

Several advantages of using embedded electrical resistance sensors were identified. In particular, such sensors allow continuous monitoring of the structure. A limitation of this type of measurement is that it is difficult to distinguish between the effects due to water and chloride ingress and those due to chemical reactions in the concrete. The Wenner four-probe surface resistivity test is an easy to use and effective tool for assessing the condition of the cover concrete. This test has similar limitations to embedded electrical sensors.

3.5 Inter-laboratory comparison

Inter-laboratory comparison is a powerful tool for evaluating the precision of test methods, as well as the performance of individual test laboratories. Owing to the fact that an inter-laboratory comparison test is both costly and time consuming, only a limited number of such tests have been done in the past decade to evaluate the precision of the test methods used to determine the resistance of concrete to chloride ingress.

3.5.1 Nordic inter-laboratory comparison

In a Nordic inter-laboratory comparison the following three Nordic test methods were evaluated (Tang and Sørensen, 1998, 2001):

- NT BUILD 355 – steady-state migration test
- NT BUILD 443 – immersion test
- NT BUILD 492 – RCM test as a non-steady-state migration test.

Nine laboratories from Nordic countries participated in the inter-laboratory comparison. Three types of concrete containing Portland cement, silica fume and ground granulate blast-furnace slag as the binder (Table 3.5) were used in the investigation. The concrete specimens were cured in water at 20°C and tested at age 5 weeks for the immersion test and 2 months for the migration tests. The later age testing schedule for the migration tests was considered in order to reach a similar concrete age in both test types, as the immersion test requires a duration of immersion of 35 days.

3.5.2 European inter-laboratory comparison (CHLORTEST)

In the European project CHLORTEST, two inter-laboratory comparison tests were undertaken, one for the preliminary evaluation of six test methods (Castellote and Andrade, 2005), and the other for the final evaluation of four test methods selected after the preliminary evaluation (Tang, 2005).

The six test methods selected for the preliminary evaluation are listed in Table 3.6. Eight laboratories from seven European countries participated in

Table 3.5 Concrete mix proportions used in the Nordic inter-laboratory comparison (Tang and Sørensen, 2001)

	Concrete		
	SF040	PC050	SL050
Cement type	CEM I 42.5 N BV/ SR/LA	CEM I 42.5 N V/ SR/LA	CEM III/B 42.5 LH HS (70% slag)
Cement (kg/m³)	386	380	390
Silica fume (slurry) (kg/m³)	34	—	—
Aggregate (kg/m³)	1860	1860	1860
Water (kg/m³)	168	190	195
Water/binder ratio	0.4	0.5	0.5
Superplasticiser (wt% of binder)	0.8	—	—
Compressive strength 28 days[1] (MPa)	82.6	63.2	45.1

Note
1 According to the Swedish Standard SS 13 72 10.

Table 3.6 Test methods selected for pre-evaluation in the European project CHLORTEST

Method	Specimen size (mm)	Pre-conditioning	Chloride exposure and measurement	Test results
EN 13396 (2004)	Ø: > 100 L: > 100[1]	Vacuum saturation and weighing every 4 hours until constant weight[2]	Immersion in 3% NaCl at 40°C for 28 days, 3 months and 6 months; chloride content measured at depths 1–2, 4–6 and 8–10 mm, respectively	Chloride content at different depths and immersion periods
NT BUILD 443 (1995)	Ø: > 70 L: > 60	Immersion in limewater until constant weight	Immersion in 165% NaCl at 23°C for > not 35 days; chloride profiles measured at less than eight depths	Curve-fitted surface chloride C_s, diffusivity D_a
NT BUILD 492 (1999)	Ø: 100 L: 50	Vacuum saturation similar to ASTM C1202, but with limewater	Migration with 10% NaCl at about 20°C under 10–60 V DC for 24 hours; chloride penetration depth and, optionally, surface chloride content measured	Diffusivity D_{nssm} and, optionally, C_s
INSA steady-state migration (Truc et al. 2000)	Ø: 100 L: 20	Vacuum saturation similar to ASTM C1202, but with demineralised water	Migration with 1 M NaCl at about 20°C under 12 V DC; chloride concentration in the upstream cell measured until a constant level observed	Diffusivity D_e
IETcc steady-state migration, (Castellote et al., 2001b)	Ø: 100 L: 20	Similar to the above	Similar to the above, but solution conductivity measured in the downstream cell	Diffusivity D_e and D_a
Resistivity (Andrade et al., 2000b)	Any	Similar to the above	Ohmic resistance of specimen measured	Resistivity

Notes

1 $L = 60$ mm was used in the evaluation in order to reduce the number of specimens required.
2 To remove the need to weigh the sample every 4 hours, a saturation procedure similar to that in ASTM C1202, but with demineralised water, was used.

the preliminary evaluation. Four types of concrete manufactured in four different European countries were used in the evaluation (Table 3.7). The preliminary evaluation found that the immersion test EN 13396 did not supply sufficient information about chloride transport, and that the measurement uncertainty in the INSA steady-state migration test was relatively large due to small changes in chloride concentration in the upstream cell. Therefore, the following four test methods were selected for the final evaluation:

1   the immersion test (NT BUILD 443), as for non-steady-state diffusion;
2   the RCM test (NT BUILD 492), as for non-steady-state migration;
3   the IETcc test (steady-state migration test based on the measurement of conductivity), as for steady-state and non-steady-state migration;
4   the resistivity test, as for the rapid indication of the transport properties.

A total of 17 laboratories from ten European countries participated in the final evaluation, in which six types of concrete with four types of binder were used. The mix proportions of the concretes tested are listed in Table 3.8. Swedish natural sand and gravel were used as fine and coarse aggregates, respectively. The concrete slabs were cast and cured under moist

Table 3.7 Mix proportions of concretes used in the CHLORTEST preliminary evaluation

	Concrete			
	SF	*PC*	*FA*	*SL*
Cast by	CHALMERS	IETcc	LNEC	TNO
Cement type	CEM I 42.5 N V/SR/LA	CEM I 42.5 R/SR	CEM IV/B 32.5 R	CEM III/B 42.5 LH HS
Cement (kg/m³)	399	400	340	350
Silica fume (kg/m³)	21 (slurry)	—	—	—
Fly ash (kg/m³)	—	—	—	—
Slag (kg/m³)	—	—	—	—
Water (kg/m³)	168	180	153	157.5
Sand (kg/m³)	842.5 (0–8 mm)	742 (0–6 mm)	62 (0–2 mm) 603 (0–4 mm)	70 (0–1 mm) 790 (0–4 mm)
Coarse aggregate (kg/m³)	842.5 (8–16 mm)	1030 (6–16 mm)	619 (4–12 mm) 555 (12–25 mm)	1040 (4–16 mm)
Total aggregate (kg/m³)	1685	1772	1823	1830 (1900?)
Super-plasticisers (kg/m³)	3.4 Cementa 92M	4.8 Melcret 222	4.1 Rheobuild 1000	3.9 Cretoplast
Air content	6%	—	—	1.5
Water/cement	0.42	0.45	—	—
Water/binder	0.40	0.45	0.45	0.45
Strength (MPa)	63	45	52.6	—
Slump (mm)	—	> 150	—	—

Table 3.8 Mix proportions of the concretes used in the CHLORTEST final evaluation

	Concrete					
	PC50	PC42	PC35	SF42	FA42	SL42
Cement type	Swedish CEM I 42.5 N BV/SR/LA				Norwegian CEM II/A-V 42.5 R (~18% fly ash)	Dutch CEM III/B 42.5 LH HS (~70% slag)
Cement (kg/m³)	400	420	450	389.5	410	410
Silica fume (ELKEM) (kg/m³)	—	—	—	20.5	—	—
Water (kg/m³)	200	176.4	157.5	172.2	172.2	172.2
Sand (0–8 mm) (kg/m³)	920	926	904	897	901	901
Gravel (10–15 mm) (kg/m³)	816	855	904	897	901	901
Superplasticiser (CemFlux) (% of binder)		0.5	1.0	0.5	0.5	0.5
Water/binder	0.5	0.42	0.35	0.42	0.42	0.42

conditions (covered with thick plastic foils) at room temperature (approximately 20°C) for at least 4 weeks; even though the slabs remained covered by the thick plastic foils, after this time the moist conditions could not be assured. After about 5 months of storage in the laboratory, cores of different sizes, as needed for the tests, were cut from the slabs. At a concrete age of about 6 months, three cores of concrete type were randomly selected and distributed to each participating laboratory for testing.

3.5.3 *International inter-laboratory comparison (RILEM)*

In parallel to the CHLORTEST preliminary evaluation, an international inter-laboratory comparison was organised by RILEM TC 178-TMC Testing and Modelling Chloride Penetration in Concrete (Castellote and Andrade, 2006). Some 27 laboratories from around the world participated in this inter-laboratory comparison for evaluating 13 test methods, including those evaluated in the CHLORTEST project (except EN 13396). It should be noted that participation in this inter-laboratory comparison was on a voluntary basis. Each participating laboratory decided itself which methods it would use. Therefore, some methods were tested by a large number of laboratories, while others were tested by only a very small number of laboratories.

Four types of concrete used in the CHLORTEST preliminary evaluation (see Table 3.7) were also used in the RILEM inter-laboratory comparison.

3.6 Precision of the laboratory test methods

The test results from individual laboratories participating in the inter-laboratory comparisons mentioned in the Section 3.5 were analysed statistically to determine the precision of each test method evaluated. The precision analysis was carried out according to ISO 5725–2:1994.

3.6.1 Results of the Nordic inter-laboratory comparison

The results of the Nordic inter-laboratory comparison are summarised in Table 3.9, and the detailed data can be found in the report by Tang and Sørensen (1998). The results show that of the values of the diffusion coefficient obtained using the RCM test (NT BUILD 492) and the natural immersion test (NT BUILD 443) are fairly comparable, although the former gives slightly higher values than the latter. However, the values of the diffusion coefficient obtained using the steady-state migration test (NT BUILD 355) are one to two orders of magnitude lower than those measured by the other two methods. Such a large difference cannot be simply explained by Eq. 2.15.

The results also show that the RCM test exhibits quite good precision, with a repeatability coefficient of variation (COV) of 5–9% and a reproducibility COV of 12–24%. The immersion test also exhibits good precision, with a repeatability COV in the range 8–14% and a reproducibility COV in the range 16–23%. The latter is comparable with the RCM test.

Table 3.9 Results from the Nordic inter-laboratory comparison

	Concrete			
	SF040	PC050	SL050	
	General mean D ($\times10^{-12}$ m²/s)			Final number of laboratories
NT BUILD 355	0.109	1.42	0.0612	2–3
NT BUILD 443	2.73	14.7	2.05	2–5
NT BUILD 492	3.11	18.6	2.55	6
	Repeatability COV(%)			Mean COV (%)
NT BUILD 355	16.6	8.2	3.5	9.4
NT BUILD 443	13	14	7.7	11.6
NT BUILD 492	8.9	5.1	8.5	7.5
	Reproducibility COV(%)			Mean COV (%)
NT BUILD 355	97.9	16.9	53.6	56.1
NT BUILD 443	15.7	22.5	18.4	18.9
NT BUILD 492	11.9	13.1	23.6	16.2

The steady-state migration test exhibits good repeatability (COV 4–17%) but very poor reproducibility, especially for concrete with pozzolanic additions (SF040 and SL050, with a COV of 98% and 54%, respectively). The test duration for these types of concrete is usually very long (1–2 months in this study). Therefore, some unintended events (e.g. leakage, change of ions in the pore solution, change of pore structures) might occur during this long time, resulting in completely different chloride flow properties, and indeed different laboratories measured rather different slopes (Figure 3.20).

3.6.2 Results of the European inter-laboratory comparison (CHLORTEST)

The results of the European inter-laboratory comparison (the CHLORTEST project) for the four selected test methods are summarised in Tables 3.10 and 3.11. The results for the other test methods and the detailed data are available in the reports by Castellote and Andrade (2005) and by Tang (2005), respectively. The results show that, of the methods for testing diffusivity, the RCM test (NT BUILD 492) exhibited the best precision (mean repeatability COV 15% and mean reproducibility COV 24%), while the steady-state migration test exhibited relatively poor precision in terms of reproducibility (mean COV 78% for the steady-state diffusion coefficient).

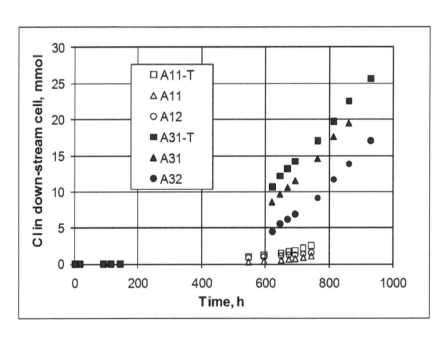

Figure 3.20 Accumulated chloride passed through concrete SF040 (data for A11-T, A11 and A12 were measured in one laboratory, and the data for A31-T, A31 and A32 in another laboratory) (Tang and Sørensen, 1998).

Table 3.10 Results of the pre-evaluation stage of the CHLORTEST project[1]

	Concrete				
	SF040	PC045	FA050	SL050	
	General mean D ($\times 10^{-12}$ m^2/s)				Final number of laboratories
NT BUILD 443 Ponding test	2.04	17.6	7.43	2.6	6–8
NT BUILD 492	2.61	15.8	5.4	2.51	7–8
IETcc, D_a	5.11	17.8	8.19	2.67	5–7
IETcc, D_e	1.16	2.9	1.01	0.45	6–7
INSA, D_e	0.997	1.85	1.04	0.994	5–7
Resistivity (Ω m)	332	70	273	382	6–7
	Repeatability COV (%)				Mean COV (%)
NT BUILD 443 Ponding test	25.1	28.1	36.5	26.5	29.1
NT BUILD 492	17.6	12.9	20.1	18.4	17.3
IETcc, D_a	21.1	13.1	25.8	34.7	23.7
IETcc, D_e	25.5	14.6	30.2	12.9	20.8
INSA, D_e	20.7	15.7	26.3	12	18.7
Resistivity (Ω m)	9.2	12.6	7.9	7.3	9.3
	Reproducibility COV (%)				Mean COV (%)
NT BUILD 443 Ponding test	47.6	41.5	44.5	35	42.2
NT BUILD 492	24.3	12.9	33.8	21.1	23.0
IETcc, D_a	40.5	39	58	43.9	45.4
IETcc, D_e	90	69	91.3	69.4	79.9
INSA, D_e	63.3	100.5	74.8	110	87.2
Resistivity (Ω m)	26.5	21.9	27.1	17.4	23.2

Note
1 Data from Castellote and Andrade (2005).

The corresponding figure for the non-steady-state diffusion coefficient was 87%. These large variations in COV are considered to arise for reasons similar to those discussed in Section 3.6.1.

The values of the non-steady-state diffusion coefficient obtained from the three different test methods for the concretes with a water/binder ratio higher than 0.45 are within in a comparable range, but the values obtained with the IETcc test are higher than those obtained from both the immersion test and the RCM test for the concretes with a water/binder ratio lower than 0.42, especially for the concrete with pozzolanic additions. The values of the steady-state diffusion coefficient are lower than those of non-steady-state one, but the difference is not as great as that found in the Nordic inter-laboratory comparison. The difference can be explained by the Eq. 2.15, and is due to a combined effect of chloride binding and the porosity of

Table 3.11 Results from the final evaluation stage of the CHLORTEST project[1]

	Concrete						
	PC050	PC042	PC035	SF042	FA042	SL042	
	General mean D ($\times 10^{-12}$ m2/s)						Final number of laboratories
NT BUILD 443	17.5	15.7	4.66	4.07	1.46	1.50	6–8
NT BUILD 492	15.4	14.7	5.94	7.44	2.31	1.77	12–13
IETcc, D_a	20.0	19.4	11.1	14.7	6.08	12.2	10–11
IETcc, D_e	2.15	1.79	1.07	1.20	0.60	0.42	10–11
Resistivity (Ω m)	54.4	60.7	124	155	323	555	7–8
	Repeatability COV (%)						Mean COV (%)
NT BUILD 443	23.2	31.4	12.5	11.3	15.5	26.6	20.1
NT BUILD 492	8.7	13.4	12.4	12.1	22.2	22.3	15.2
IETcc, D_a	28.8	27.6	17.8	51.2	50.7	38.2	35.7
IETcc, D_e	30.2	27.4	22	25.7	19.8	22.3	24.6
Resistivity (Ω m)	6.6	10.6	7.6	14.5	10.1	13.7	10.5
	Reproducibility COV (%)						Mean COV (%)
NT BUILD 443	32.3	35.6	26	26.7	19.3	30	28.3
NT BUILD 492	16.7	21.9	16.6	21.6	31.2	33.4	23.6
IETcc, D_a	96.3	77.8	77.4	79.3	93.2	99.4	87.2
IETcc, D_e	84.4	84.8	71.5	67.9	89.8	69.1	77.9
Resistivity (Ω m)	18.9	15.1	21.1	26.9	33.6	35	25.1

Note
1 Data from Tang (2005).

concrete. The difference probably arises because of the thinner specimen used (thickness 20 mm in the IETcc test compared with 50 mm in the NT BUILD 355 test), which allows for earlier establishment of the steady-state condition.

3.6.3 Results of the international inter-laboratory comparison (RILEM)

The results of the international inter-laboratory comparison (RILEM TC 178-TMC) for the four selected test methods are summarised in Table 3.12. The results for the other test methods and the detailed data are available in the report by Castellote and Andrade (2006). Again, the results show that the RCM test (NT BUILD 492) exhibited the best precision (mean repeatability COV 18% and mean reproducibility COV 36%), although the reproducibility COV is higher than those obtained in the Nordic and the European evaluations. This is probably due to factors such as the involvement of some inexperienced laboratories in the programme and the relatively large variations in testing periods between different laboratories. The steady-state migration test also exhibited relatively poor precision in terms of the

Table 3.12 Results from the RILEM inter-laboratory comparison[1]

	Concrete				
	SF040	*PC045*	*FA050*	*SL050*	
	General mean D *(× 10^{-12} m2/s)*				*Final number of laboratories*
NT BUILD 443	2.21	19.4	7.12	2.63	11–16
Ponding test	3.03	20.9	7.3	2.58	3–11
NT BUILD 492	2.35	12.6	5.27	2.46	13–17
IETcc, D_a	5.29	15.6	7.13	2.83	7–9
IETcc, D_e	0.891	2.23	0.668	0.494	9–11
INSA, D_e	0.748	1.12	0.957	0.552	7–8
Resistivity (Ω m)	314	72.5	257	367	10–12
	Repeatability COV (%)				*Mean COV (%)*
NT BUILD 443	26.9	28.5	30.8	25	27.8
Ponding test	17.7	26.2	27.3	21	23.1
NT BUILD 492	18.6	13.4	19.8	20.9	18.2
IETcc, D_a	24.7	22	25	29.5	25.3
IETcc, D_e	27.6	16.6	34.3	11.1	22.4
INSA, D_e	14.4	16.7	26.4	19.3	19.2
Resistivity (Ω m)	8	12	11	7	9.5
	Reproducibility COV (%)				*Mean COV (%)*
NT BUILD 443	45.3	52.7	45.9	45.7	47.4
Ponding test	72.6	71.5	54.2	–[2]	66.1
NT BUILD 492	36.6	37.1	36.9	33.9	36.1
IETcc, D_a	47.3	49.2	61.2	37.2	48.7
IETcc, D_e	67.6	65.1	83	61	69.2
INSA, D_e	71.3	68.8	70.6	95.9	76.7
Resistivity (Ω m)	27	29	25	24	26.3

Notes
1 Data from Castellote and Andrade (2006).
2 Value ignored due to the incredibly low value, which is less than that of repeatability.

reproducibility (mean COV 69%). The corresponding value for the non-steady-state migration was 49%. These large variations are considered to have arisen for be reasons similar to those discussed in Section 3.6.1.

The overall mean values of the test results from the RILEM evaluation are closely comparable to those obtained in the CHLORTEST pre-evaluation, because the same types of concrete were used in both evaluations.

3.6.4 *Summary of the precision results*

The precision results obtained in the different inter-laboratory comparisons are summarised in Figures 3.21 and 3.22. It can be seen that the precision

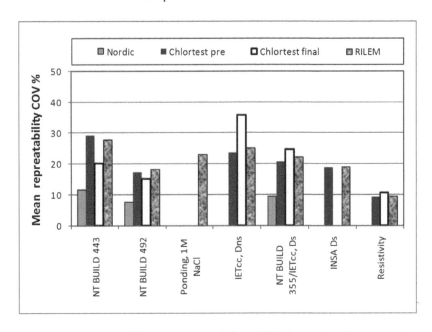

Figure 3.21 Summary of the mean repeatability COV obtained in the various inter-laboratory comparisons.

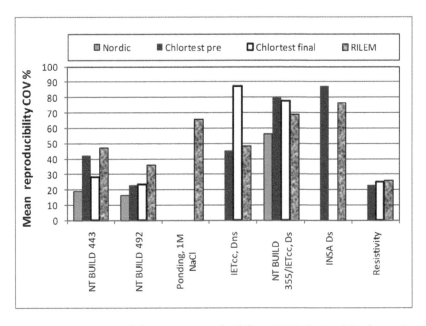

Figure 3.22 Summary of the mean reproducibility COV obtained in the various inter-laboratory comparisons.

of each test method is, in general, dependent on the number and type of organisations involved in the inter-laboratory comparison, except for the resistivity test, which is a relatively simple to measure. This is understandable, because many test methods require laboratory experience to reduce the variability of test results. In principle, prior to an inter-laboratory comparison, the experience of the laboratories in the specified test methods should be examined. Unfortunately, there was rarely an opportunity to carry out such examinations, due to limitations of finance and time, especially when the participation in the inter-laboratory comparison was on a voluntary basis. From the viewpoint of the statistical base, an inter-laboratory comparison requires a large number of participating laboratories. However, on the other hand, a large number of participating laboratories also increases the variation in the quality of the participants. Therefore, in practical terms, an inter-laboratory comparison involves a compromise between the number of participating laboratories and laboratory quality. The inter-laboratory comparisons organised by the CHLORTEST project and RILEM TC 178-TMC involved similar descriptions of test methods and comparable numbers of participating laboratories. It is, therefore, reasonable to take the precision values determined in the CHLORTEST and RILEM inter-laboratory comparisons to represent the precision of each test method, as listed in Table 3.13.

In summary, the RCM test (NT BUILD 492) has the best precision of all the diffusion/migration tests evaluated. The ponding test shows a relatively lower precision than the immersion test NT BUILD 443. Both the IETcc and INSA steady-state migration tests show relatively poor reproducibility, although their repeatability is comparable with that of the immersion test. The resistivity test shows even better precision than the RCM test, but a calibration is required in order to convert the results to chloride diffusivity. Calibration of any method will introduce uncertainty, and the final precision is related to the sum of the squared standard deviations for both the resistivity test and the method of calibration.

Table 3.13 Summary of the precision of methods for testing chloride resistance

Method	Parameter	Mean repeatability COV (%)	Mean reproducibility COV (%)
NT BUILD 443	D_{nss}	25.6	39.3
NT BUILD 492	D_{nss}	16.9	27.6
Ponding with 1 M NaCl	D_{nss}	23.1	66.1
Steady-state migration test (IETcc test)	D_{ss}	28.2	60.4
	D_{nss}	22.6	75.7
Steady-state migration test (INSA test)	D_{ss}	18.9	81.9
Resistivity test	ρ	9.8	24.9

3.7 Relationships between the results of the different test methods

3.7.1 Effect of concrete age on the test results

As the properties of concrete are age dependent, especially for young concrete, results obtained using different test methods should be compared using concrete of a similar age. In the inter-laboratory comparisons described in Section 3.6, the effect of concrete age was considered by starting the test at different ages (in the Nordic comparison) or at a concrete age of more than 6 months (in the CHLORTEST and RILEM comparisons), when the age was significantly greater than the test duration. When comparing results reported in the literature, this age effect should be taken into account.

Owing to the rapidity of the RCM test, the age effect can be investigated easily. Tang (1996a) has presented some test results for Portland cement and silica fume concrete, and these are shown in Figure 3.23. It can be seen that the diffusion coefficient decreases with concrete age when the concrete is young (less than half a year in most cases), and is dependent on the water/binder ratio and the type of binder. Based on these results, Tang and Sørensen (1998) proposed the following equation to correct the test results for age for Portland cement concrete:

$$\frac{D_{nssm}(t_2)}{D_{nssm}(t_1)} = \left(\frac{t_1}{t_2}\right)^{0.152\left(w/_c\right)^{-0.6}} \tag{3.17}$$

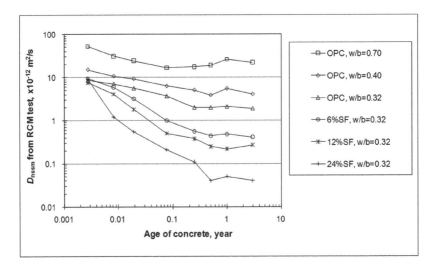

Figure 3.23 Effect of concrete age on the non-steady-state diffusion coefficient. OPC, ordinary Portland cement; SF, silica fume; w/b, water/binder. (Based on Tang (1996a).)

3.7.2 Relationship between the results of the diffusion and migration tests

As the immersion test (NT BUILD 443) and the ponding test are close to the real situation in all the methods evaluated in the inter-laboratory comparisons, the values obtained using this method could be used as reference values for comparison. Although the chloride concentration and the duration of exposure in NT BUILD 443 (165 g NaCl per litre for 35 days) and the ponding test (1 M = 58.45 g NaCl per litre for 90 days in the RILEM inter-laboratory comparison) are different, the overall mean values obtained with these two methods are closely comparable (Figure 3.24). As the diffusion coefficient on the one hand increases with decreased concentration, and that on the other hand decreases with increasing exposure time, the two effects compensate for each other in the immersion and ponding tests, resulting in comparable test values. Taking into account the shorter exposure time and relatively better precision of the NT BUILD 443 test, the results obtained using this method are compared with those obtained in the migration tests in Figures 3.25 and 3.26.

It can be seen from Figure 3.25 that the diffusion coefficients determined using the RCM test and the immersion test are fairly comparable in most cases. Obviously, most of the measurements are in the range of $\pm s$, where s is the standard deviation of the reproducibility (see Table 3.13) of the NT BUILD 443 test (x axis) and the NT BUILD 492 test (y axis). The data

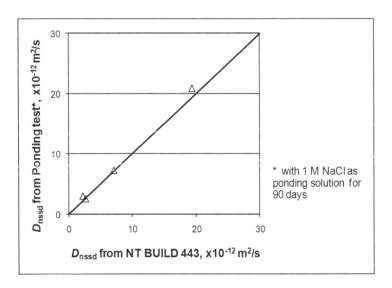

Figure 3.24 Relationship between the results obtained in the NT BUILD 443 immersion test and the ponding test (1 M NaCl ponding solution for 90 days), with different concentrations and exposure time. (Data from Castellote and Andrade (2006).)

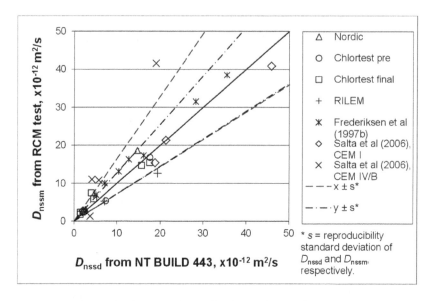

Figure 3.25 Relationship between the results obtained using the immersion test and the RCM test. *s* = standard deviation of the reproducibility of D_{nssd} and D_{nssm}.

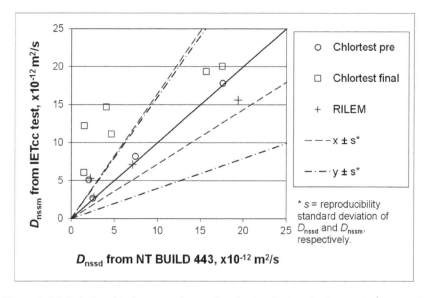

Figure 3.26 Relationship between the results obtained using the immersion test and the D_{nss} values obtained using the IETcc steady-state migration test. *s* = standard deviation of the reproducibility of D_{nssd} and D_{nssm}.

from the fly ash cement type CEM IV/B, reported by Salta *et al.* (2006), are outside the range ±*s*. A possible reason for this could be the age effect, because the age of the concrete at the start of these tests was 28 days, while under the exposure duration in the NT BUILD 443 test the hydration of fly ash cement can increase significantly, resulting in a lower diffusivity.

Figure 3.26 shows that the values of D_{nssm} obtained from the IETcc test are reasonably comparable with those obtained with the NT BUILD 443 test for the relatively porous concrete samples, but deviate for the dense concrete samples (low water/binder ratio, or having pozzolanic additions). As discussed previously, the defects, if any, in a specimen or any leakage from the curved surface of the specimen may cause a shorter time lag, which, in turn, results in a larger calculated diffusion coefficient.

Figure 3.27 shows the relationship between the values obtained from the two steady-state migration tests, the IETcc and INSA tests. Despite the fact that both methods exhibit relatively poor reproducibility, the values are within a comparable range, although the linear relationship is not clear.

The relationship between the values obtained from the immersion test and the steady-state migration tests is shown in Figure 3.28. As expected, the steady-state diffusion coefficient is smaller than the non-steady-state one at all times. This is mainly due to the fact that the effective diffusible area (porosity) is less than the apparent cross-sectional area of a concrete specimen, while the effect of chloride binding counteracts this difference (see Eq. 2.15).

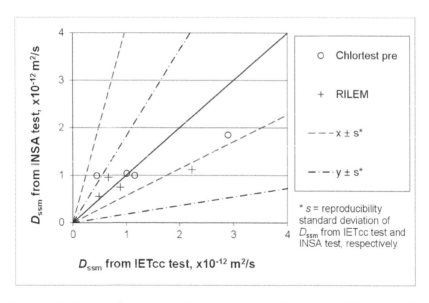

Figure 3.27 Relationship between the results obtained using the IETcc and INSA steady-state migration tests. *s* = standard deviation of D_{ssm}.

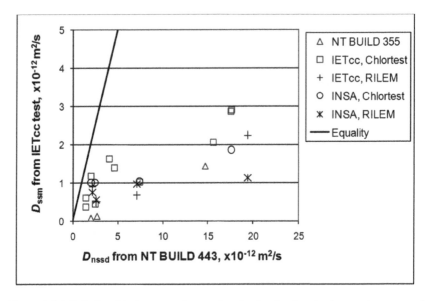

Figure 3.28 Relationship between the results obtained using the immersion test and the D_{ss} value obtained using the steady-state migration test.

3.7.3 Relationship between the results of the resistivity and diffusion/migration tests

According to Eq. 3.14, if the chloride transference number is known, the resistivity ρ can simply be converted to the effective or steady-state diffusion coefficient D_{ss}:

$$D_{ss} = \frac{k_{Cl}}{\rho} \tag{3.18}$$

where k_{Cl} becomes a constant. According to Andrade *et al.* (2000b), when the dimensions of D_{ss} are cm²/s and those of ρ are Ωcm, the value of k_{Cl} is $1.2 \times 10^{-4}\,\Omega\,\text{cm}^3/\text{s}$, corresponding to $120\,\Omega\,\text{m}^3/\text{s}$ or $0.12\,\text{k}\Omega\,\text{m}^3/\text{s}$.

Figure 3.29 shows the relationship between resistivity (conductivity as its inversion) and the steady-state migration coefficient. It can be seen that the value of k varies from 0.1 to $0.2\,\text{k}\Omega\,\text{m}^3/\text{s}$, depending on the test series. As the precision of the resistivity test is apparently better than that of the steady-state migration test, the variation in the k value can be attributed to both the inconstant chloride transference number and the inaccuracy in the D_{ss} measurement.

The relationships between resistivity and non-steady-state diffusion/migration coefficient are shown in Figures 3.30 and 3.31. It can be seen that the k

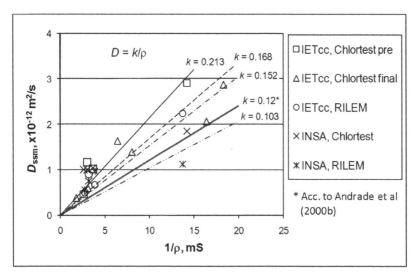

Figure 3.29 Relationship between the results obtained with resistivity test and the steady-state migration test. *Data from Andrade *et al.* (2000b).

values vary even more than that in Figure 3.29, indicating that the correlation between resistivity and diffusion coefficient is case dependent and that there is no general constant k value to convert the values from the resistivity test to the diffusivity tests, as established in Section 3.3.6.

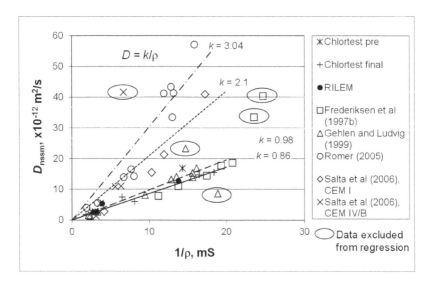

Figure 3.30 Relationship between the results obtained with the resistivity test and the non-steady-state migration test.

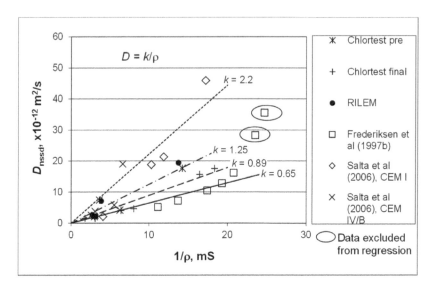

Figure 3.31 Relationship between the results obtained using the resistivity test and the immersion test.

3.7.4 Relationship between laboratory tests and in-field performance

When dealing with the durability of concrete, long-term experiments and observations are required to establish a relationship between the results of laboratory tests and the measured in-field performance. Tang (2003a) investigated chloride ingress in over 30 types of concrete in a marine environment for over 10 years, and found a reasonably good relationship between the migration coefficient obtained from the NT BUILD 492 test at a concrete age of 6 months and the apparent diffusion coefficient curve-fitted from the field data, as shown in Figure 3.32. However, this relationship is time dependent, i.e. it changes with the exposure period, as shown in Figure 3.33. This is because of the time-dependent behaviour of the apparent diffusion coefficient, which is discussed in more detail in Chapter 4. Despite this time-dependent behaviour, the migration coefficient measured in the laboratory reflects the chloride ingress in the field fairly well, as shown in Figure 3.32. An investigation of concrete in a road environment (Tang and Utgenannt, 2007) has also confirmed the excellent relationship between the migration coefficient obtained with the NT BUILD 492 test and the apparent diffusion coefficient curve-fitted from the field data, as shown in Figure 3.34.

Salta *et al.* (2006) compared the laboratory test data and in-field data after 3 years exposure, and concluded that the diffusion or migration coefficient

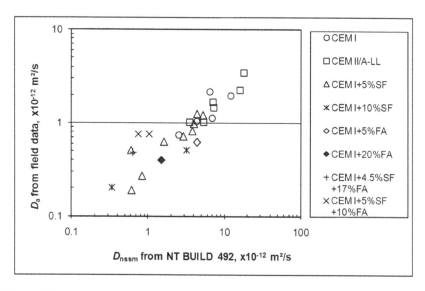

Figure 3.32 Relationship between the results obtained in the laboratory RCM test at 6 months and the apparent *D* measured in the field after 10 years exposure (Tang, 2003a). FA, fly ash; SF, silica fume.

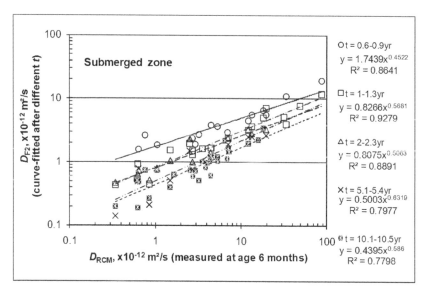

Figure 3.33 Relationship between the results obtained in the laboratory RCM test at 6 months and the apparent *D* measured in the field after different exposure periods (Tang, 2003a).

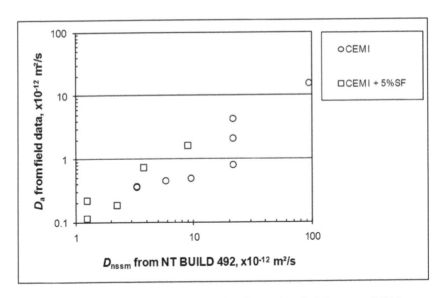

Figure 3.34 Relationship between the results obtained in the laboratory RCM test at
6 months and the apparent *D* measured in the field after 10 years exposure in a road environment (samples taken from the vertical surfaces)
(Tang and Utgenannt, 2007). SF, silica fume.

obtained in the laboratory using either the NT BUILD 443 or NT BUILD 492
test at a concrete age of 28 days can be used for concrete ranking. They
also concluded that such data can be used to define durability indicators, to
ensure the long-term durability of concrete structures.

4 Modelling of chloride ingress

4.1 Introduction

A model for predicting chloride ingress into concrete at all times aims at predicting the chloride profile $C(x, t)$ after a certain exposure time t or at least the chloride content C at the depth of the reinforcement. The output is usually intended to be compared with a 'chloride threshold level', which is relevant for reinforcement corrosion (Figure 4.1).

Any prediction model should have some input data that includes information on the concrete and information on the environment. A model can be considered as the process of obtaining either a predicted chloride profile or the chloride content from the input data. The output should then fit or explain performance data obtained from the field in various environments. Examples of such environments (and relevant structures) are: submerged, tidal/splash, atmospheric, wicking (tunnels, caissons), de-icing (roads, road bridges, parking decks, stairs), and pools.

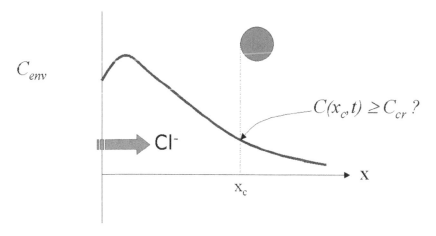

Figure 4.1 The comparison between the predicted chloride content at the depth of the reinforcement and the chloride threshold level for corrosion.

4.2 Principles of the ingress process

4.2.1 Concrete as an ingress medium

Chloride-ingress models are generally applied to homogeneous, crack-free concrete with a one-dimensional ingress at a macro-scale. Some exceptions can be found in literature, but the main difference is in geometrical considerations. The analysis in this chapter is limited to one-dimensional ingress models for a crack-free concrete, although a one-dimensional model might be applied to two- or three-dimensional cases, possibly with cracks, by means of pure mathematics and geometry.

The assumption of homogeneity for concrete, however, might be too simplistic, even if the concrete is very well mixed and compacted. The wall effect will certainly create a binder content profile that is closer to a cast concrete surface. This will directly influence the shape of the chloride-ingress profile and the total chloride contents. This is particularly significant if the depth of penetration of chlorides is small because the influencing depth of a large wall effect is at least half of the maximum size of the coarse aggregate. Furthermore, at greater depths the wall effect is influenced by the aggregate content and, consequently, the binder content.

Vertical separation may give differences in the water/cement ratio and binder content at different vertical levels. Concrete will change over time, and these changes are somewhat different at different depths. The continuous binder reactions will densify concrete over time and, consequently, change the pore system over time. Changes with depth will depend on the initial curing and the moisture conditions of the concrete created by the exposure conditions during its service life.

Concrete will also interact with the surrounding environment in other ways which will change the material over time and in different ways at different depths. Examples of mechanisms that will have such an effect, and influence the chloride-ingress process, are drying and wetting, which cause shrinkage and swelling, and carbonation.

4.2.2 The concrete pore solution

All types of concrete undergo 'self-desiccation' due to the reduction in the amount of chemically bound water. This means that the pore system is not saturated. For low water/cement ratio concretes, this also means that the initial pore water has an activity, a relative humidity (RH), much lower than 100%. Concretes intended for use in chloride environments may undergo self-desiccation, which results in a relative humidity inside the concrete as low as 85–90%, corresponding to a degree of capillary saturation of the pores, excluding the air pores, of some 0.8–0.9.

The pore liquid is not pure water, but a strong solution containing ions, mainly sodium, potassium and hydroxyl ions, besides the chloride ions, the composition being different for different types of binder and different compositions of concrete. The self-desiccation will increase further the concentrations of these ionic species in the pore solution.

4.2.3 *The exposure conditions*

The boundary conditions for the chloride-ingress process differ widely according to the environment in which the concrete is placed. In the simplest environment, such as the submerged zone of a marine concrete structure, the main components of the seawater are chloride and sodium, but a number of other ions are present in small quantities. Depending on the local conditions, the salinity and the seawater temperature may undergo annual variations. Consequently, the exposure solution is quite different from the pore solution in the concrete, not only with regard to the chlorides, but also with regard to other ions.

In other environments the exposure conditions can be much more complicated. A 'constant' exposure solution may not be present all the time, occasionally being replaced by water from driving rain or undergoing drying conditions, as in the splash zone of marine structures. In an environment where de-icing salts are present, the exposure solution contains a large amount of sodium chloride for a short while, and is then diluted, and sometimes replaced, by water from rain. In addition, most of the time the concrete surface is exposed to drying conditions.

4.2.4 *The ingress process*

For a saturated concrete constantly exposed to a sodium chloride solution, the difference in ionic strength between the pore solution and the exposure solution will initiate the transport of ions. Chloride ions will move into the concrete pore solution and some of the alkalis and hydroxides in the pore solution will start to leach out. This is popularly expressed as 'diffusion' of ions, but various ions will interact with each other in such a way that the charge balance is maintained. The flux of one type of ion is affected by the flux of other ion types because of the electrical field created by the ions.

In concrete in a structure, the pore system is not saturated with water. Consequently, there will be moisture flow into the pore system. In addition, depending on the environmental conditions, the moisture flow can vary over time, both in magnitude and direction, leading to convection of ions in the pore system and in and out of the concrete. Below a certain moisture content, the moisture flow cannot carry any ions, and this may cause an accumulation of ions in certain areas.

The movement of ions in the pore system is accompanied by a significant interaction with the cement matrix, sometimes called 'binding', especially for chlorides and hydroxides. Due to this interaction further penetration of chloride is significantly delayed. This interaction depends mainly on the type and amount of binder in the concrete, but is affected by, for instance, the temperature and the concentration of alkalis.

The chloride interaction/binding seems to be reversible to a large extent. This means that, if the concentration of chloride decreases (e.g. if the concrete surface is temporarily exposed to rain or if the concrete temperature increases), part of the binding is lost. Carbonation or possible sulphate ingress will also cause a loss of most of the binding capacity.

In conclusion, chloride ingress is never a simple 'diffusion' process due to concentration differences alone, and the 'chloride-binding capacity' is not a simple, constant property. Chloride-ingress models must include a number of assumptions and simplifications, and these must be stated. The problem is to realise the significance of these simplifications. This is discussed further in the sections in this chapter.

4.3 'Fundamentals' of ingress models

In this section the relevant 'fundamentals' of chloride-ingress models are defined. The objective is to state what fundamentals could be agreed upon, without discussing specific models, and to have this as a background when the different models are presented, analysed and discussed in later sections.

4.3.1 Mass balance equations

The mass balance equation for chlorides can be expressed in different ways, depending on whether separate terms are used for diffusion and binding. The simplest form gives the balance of the total amount of chloride per unit volume.

The flux of chloride J_{Cl} is different at different positions in the concrete. The difference in chloride flux to and from an infinitesimal slice of concrete with a thickness dx, as shown in Figure 4.2, will change the total amount of chloride ions C in such a slice. The change in total chloride content per

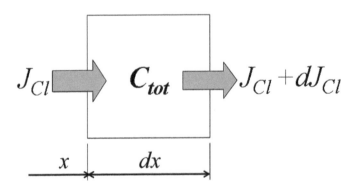

Figure 4.2 A visual description of the mass balance equation for chlorides.

unit of time will be equal to the difference in chloride flux to and from the slice, divided by the thickness of the slice. Consequently, the mass balance equation will be:

$$\frac{\partial C}{\partial t} = \frac{\partial c_t}{\partial t} = -\frac{\partial J_{Cl}}{\partial x} \tag{4.1}$$

To get the dimensions correct, the chloride content c_t here is the content per unit volume of *concrete*, not the pore volume, and the flux J_{Cl} is per unit area of *concrete*. The negative sign simply indicates that the chloride content will decrease if dJ_{Cl} is positive. For chloride ingress, dJ_{Cl} is negative, and the chloride content will increase with time (see Chapter 2 for a more detailed description of dimension and diffusion functions).

The change in total chloride content can also be split into a change in free chloride dissolved in the pore water and a change in bound chloride in such a way that equilibrium will be maintained between the free and bound chloride ions. As the chloride flux occurs in the pore solution, the change in chloride content will first occur as a change in free chloride ions. An almost 'instant' change in bound chloride c_b will follow, because the rate of chloride binding is fairly high. The mass balance equation will then be:

$$\frac{\partial C}{\partial t} = \frac{\partial c}{\partial t} + \frac{\partial c_b}{\partial t} = -\frac{\partial J_{Cl}}{\partial x} \tag{4.2a}$$

or

$$\frac{\partial c}{\partial t} = -\frac{\partial J_{Cl}}{\partial x} - \frac{\partial c_b}{\partial t} \tag{4.2b}$$

4.3.2 Flux descriptions

Traditionally, Fick's first law has been used to describe the chloride flux as a result of pure diffusion in liquid water, with the free concentration c of dissolved chloride as the transport potential:

$$q_{Cl} = D_{F1} \frac{\partial c}{\partial x} \tag{4.3}$$

The 'diffusion coefficient' D_{F1} (as D_{ss} in Chapters 2 and 3) is then defined by Eq. 4.3 and by the test set-up used to measure it (see Chapters 2 and 3).

Nowadays it is widely recognised that Fick's first law is an oversimplification of chloride transport. Instead, the effects of the electric potential field Φ, due to other ions and the potential difference applied in migration tests, are included in the flux equation. The flux is then described by the Nernst–Planck equation:

$$q_{Cl} = -D_i \left(\frac{\partial c_i}{\partial x} + c_i \frac{\partial \ln \gamma_i}{\partial x} + \frac{z_i F}{RT} c_i \frac{\partial \Phi}{\partial x} \right)$$

(4.4)

The 'diffusion coefficient' in Fick's first law D_{F1} is obviously not a material property but depends on the conditions. Consequently, it cannot be determined by a simple 'diffusion' test. Any attempt to measure D_{F1} using straightforward diffusion cells will determine something else, because Eq. 4.3 is not a correct description of what happens in such a test.

Instead it must be acknowledged that the flux of chlorides is influenced by other ions. In any test set-up for determining the 'diffusion coefficient' D_i for chloride, using Eq. 4.4, the result will depend on the diffusion coefficients of all other ions. Consequently, it is not possible to determine the chloride diffusion coefficient directly in one simple test. Different approaches to solving this problem have been used. Truc (2000) made estimations from chloride migration tests where the diffusion coefficients for sodium, hydroxyl and potassium ions were adjusted to fit predictions made using a multi-species model. Samson and Marchand (1999) instead utilised the diffusion coefficients for several ions from data obtained for diffusion in a solution. By determining the 'formation factor', i.e. the effect of the tortuosity and the restrictivity of the pore system, they could describe the flux of all ions by applying the same formation factor to the flux of all ions.

The main difficulty in applying Eq. 4.4 is to quantify the potential gradient $\partial \Phi / \partial x$, because it is a function of the fluxes of all ions and, consequently, changes with time. In a migration test, simple versions of Eq. 4.4 are used because the applied external electrical field may dominate the potential field due to the chlorides present in the pore solution.

4.3.3 Interaction/binding

The interaction between chlorides and the matrix of cement-based materials is still not very well understood. Thus, for most applications, a 'binding isotherm' is used to give the relationship between the free and the bound chloride ions, as described in Section 2.3.3. Here many questions are still disputed, besides the approach of using a binding isotherm at all. As the binding isotherm is a pure empirical 'property', it must be measured, and the effect of a number of parameters must be quantified.

The *shape* of the binding isotherm is one such questions (Figure 4.3). Does most of the binding really occur at concentrations close to zero, or is the chloride binding significantly concentration dependent?

Another important question is the relationship between 'free' and 'bound' chloride ions (Figure 4.4). As discussed in Section 2.3.3, chloride binding has a non-linear behaviour and the binding capacity is dependent on the chloride concentration, but it is sometimes assumed to be a constant. As

[Bound or total chloride]

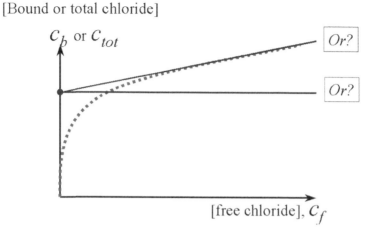

Figure 4.3 Alternative shapes of a possible 'binding isotherm'.

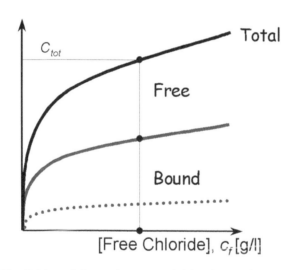

Figure 4.4 The division of the total amount of chloride into 'free' and 'bound'.

shown in Figure 2.2, the chloride-binding capacity is very high at low chloride concentrations.

Some 'strange' time effects have been observed on chloride binding in submerged concretes (Tang, 2003a,b). The surface chloride content has been found to be time-dependent $C_s(t)$. A longer exposure time gives not only deeper chloride ingress but also higher chloride content (Figure 4.5). These observations cannot be explained with current knowledge. The con-

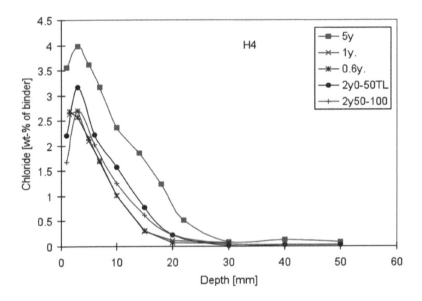

Figure 4.5 Chloride profiles after several exposure times, showing a higher chloride concentration with time. SRPC (sulphate resistant Portland cement) concrete with water/binder = 0.40 and 5% silica fume submerged for 0.6, 1, 2 and 5 years. (Data from Tang (2003b).)

sequences of alternative causes and how these findings are utilised in chloride-ingress models are quite contradictory (Figure 4.6). A time effect on the 'surface chloride content' C_s in an empirical ingress model, which has no simple explanation, has a significant consequence, i.e. a greater depth of

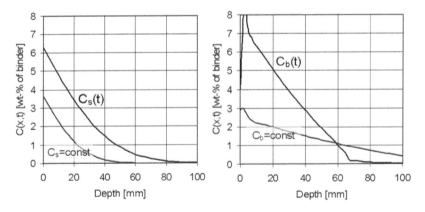

Figure 4.6 The consequences of alternative causes of the findings in Figure 4.5: (left) time-dependent surface chloride content $C_s(t)$; (right) time-dependent chloride binding $C_b(t)$.

penetration than with a constant C_s. Conversely, a time effect on chloride binding will give a smaller predicted depth of penetration, simply because the larger chloride-binding capacity with time will retard the ingress even more.

4.4 Chloride-ingress models based on Fick's second law

4.4.1 Fick's second law

What is usually called Fick's second law is in reality a mass balance equation. Originally it was meant for mass balance in solutions, where no binding exists, but it has been widely applied to mass transport in porous systems, with and without binding.

The law of mass conservation in a small volume of a solution gives the changes with time of the (free) chloride concentration in that unit volume:

$$\frac{\partial c}{\partial t} = -\frac{\partial q_{Cl}}{\partial x} \tag{4.5}$$

This is the same equation as Eq. 4.1, with the free chloride content equal to the total content, and Eq. 4.2, with the binding term equal to zero.

With the flow description as given by Fick's first law for diffusion in a solution (Eq. 4.3), and with the diffusion coefficient D_{F1} as a constant, it is possible to write:

$$\frac{\partial c}{\partial t} = D_{F1} \frac{\partial^2 c}{\partial x^2} \tag{4.6}$$

This looks like 'Fick's second law', but it should be noted that this equation is applicable only for solutions.

For a porous material, with binding, Eq. 4.2 is applicable. By assuming that the flux can be described by Fick's first law (Eq. 4.3), and that D is a constant, one obtains:

$$\frac{\partial C}{\partial t} = -\frac{\partial}{\partial x}\left(-D_{F1} \frac{\partial c}{\partial x}\right) = D_{F1} \frac{\partial^2 c}{\partial x^2} \tag{4.7}$$

This equation is not identical to 'Fick's second law' because the concentrations are different in the two terms: C is expressed per volume of concrete, but c is expressed per volume of pore solution. A factor equal to p_{sol}, according to Eq. 2.1:9, must be inserted. In addition, C is the total content but c is the free content. A factor of

$$\frac{\partial C}{\partial c} = \frac{\partial c}{\partial c} + \frac{\partial c_b}{\partial c} = 1 + \frac{\partial c_b}{\partial c} \tag{4.8}$$

must be included to correct for this. Then the mass balance equation will be:

$$\frac{\partial C}{\partial t} = D_{F1} \frac{\partial}{\partial x} \frac{\partial c}{\partial x} = \frac{D_{F1}}{p_{sol}\left(1 + \dfrac{\partial c_b}{\partial c}\right)} \frac{\partial}{\partial x} \frac{\partial C}{\partial x} = \frac{D_{F1}}{p_{sol}\left(1 + \dfrac{\partial c_b}{\partial c}\right)} \frac{\partial^2 C}{\partial^2 x} \tag{4.9}$$

Comparing this equation with the traditional Fick's second law,

$$\frac{\partial C}{\partial t} = D_{F2} \frac{\partial^2 C}{\partial^2 x} \tag{4.10}$$

it may be realised that the two equations are equal, but that the 'diffusivity' D_{F2} in Fick's second law (as D_{nss} in Chapters 2 and 3) is different from the diffusion coefficient D_{F1} in Fick's first law (as D_{ss} in Chapters 2 and 3), for a porous material with binding. Note the difference between a diffusion coefficient D and a diffusivity D. The same symbol D is used and the dimension is m²/s for both, but they are defined by Fick's first and second laws, respectively. The binding capacity is included in the diffusivity, in the same way it is included in heat conductivity, heat capacity and heat diffusivity.

In fact, the total concentration C in Fick's second law could be replaced by the free concentration c, but the diffusivity is still the same, D_{F2}:

$$\frac{\partial c}{\partial t} = D_{F2} \frac{\partial^2 c}{\partial^2 x} \tag{4.11}$$

From Eqs 4.9 and 4.10 the relationship between the two coefficients in Fick's laws is clear:

$$D_{F2} = \frac{D_{F1}}{p_{sol}\left(1 + \dfrac{\partial c_b}{\partial c}\right)} \tag{4.12}$$

which is similar to Eq. 2.15. The magnitude of the relationship depends on the porosity and the binding capacity of the concrete.

The chloride diffusion coefficients will change over time, as the concrete matures. The change with time of D_{F2} will depend on both the change in D_{F2} and the change in the binding capacity $\partial c_b/\partial c$. A few measurements have been done on the effect of ageing on D_{F2} (see the examples in Figure 3.23).

Tang (1996b) proposed an age dependence of the diffusivity measured using the rapid chloride migration (RCM) test according to Eq. 4.13, for the first 6 months:

$$\frac{D_{RCM}(t)}{D_{RCM}(t_0)} = \left(\frac{t_0}{t}\right)^{\beta} \tag{4.13}$$

where t is the age of the concrete, t_0 is the age when the diffusivity has reached a constant value (usually 6 months for OPC concrete), and β is a parameter (an 'age exponent'). In Eq. 4.13 the concrete age is limited to $t \leq t_0$.

These findings have been used to express the general age dependence of the diffusivity in Fick's second law:

$$D_{F2}(t) = D_{F2}(t_0)\left(\frac{t_0}{t}\right)^{n} \tag{4.14}$$

where the age exponent n equals β. Equation 4.14 has been extended for ages beyond t_0, for which there is no experimental evidence.

The age-dependent diffusivity in Fick's second law $D_{F2}(t)$ is also called the 'point-wise' or 'instantaneous' diffusivity.

4.4.2 Error function complement (erfc) model with constant D and C_s

This section includes the definition of the 'achieved' or 'apparent' diffusivity D_a.

The mathematical solution to Fick's second law (Eq. 4.10) is well known for a semi-infinite medium, with a constant diffusivity $D = D_{F2}$ and a constant surface chloride content $C(x=0, t) = C_s$. If the initial chloride content C_i is negligible, the solution is the complement to the error function (Crank, 1975):

$$C(x, t) = C_s \text{erfc}\left(\frac{x}{2\sqrt{Dt}}\right) \tag{4.15}$$

where t is the exposure time. This equation gives a 'chloride profile' $C(x, t)/C_s = \text{erfc}(z)$, with $z = x/(2\sqrt{Dt})$ (Figure 4.7).

Collepardi et al. (1970, 1972) were the first to use this erfc model for chloride ingress into concrete. This model then became the dominant model for almost 25 years.

The apparent diffusivity D_a and the apparent surface chloride content C_{sa}

Frequently, the erfc solution to Fick's second law (Eq. 4.15) is fitted to chloride profiles measured from structures or specimens. In such a curve-fitting, if the exposure time t is inserted, the best curve-fitting gives two regression

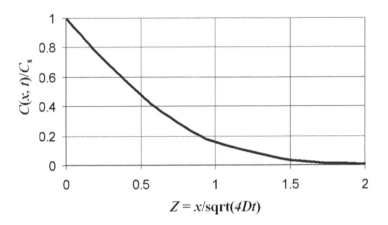

Figure 4.7 A chloride profile. Chloride concentration with 'normalised' depth, from the erfc solution to Fick's second law.

parameters: D_a and C_{sa}. The index 'a' means 'achieved' or 'apparent'. The definitions are:

- D_a is the 'apparent diffusivity' achieved after an exposure time t, assuming that the diffusivity D_{F2} in Fick's second law was constant during the whole exposure. Then $D_a = D_{F2}$.
- C_{sa} is the 'apparent surface chloride content' achieved after an exposure time t, assuming that the boundary condition was constant during the exposure. Then $C_{sa} = C(x = 0, t)$.

Solutions to Fick's second law for a limited thickness

The erfc solution to Fick's second law is applicable only for semi-infinite conditions. When the sample or structure thickness is limited, the solution is quite different (Crank, 1975):

$$C(x, F_0) = C_s U(x / L, F_0) =$$

$$C_s \left[1 - \frac{4}{\pi} \sum_{n=0}^{\infty} \frac{-1^n}{2n+1} \exp\left(-(2n+1)^2 \frac{\pi^2 F_0}{4}\right) \cos\left(\frac{(2n+1)\pi}{2} \frac{L-x}{L}\right) \right] \quad (4.16)$$

where the Fourier number F_0 is equal to $F_0 = D_a t/L^2$. L is the thickness of a structure or specimen with one-sided chloride ingress, or half the thickness in case of two-sided ingress. Equation 4.16 is shown graphically in Figure 4.8.

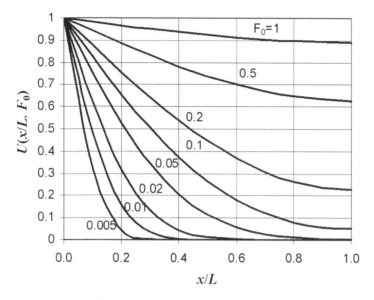

Figure 4.8 Chloride profiles for cases where the thickness is limited, obtained from the solution to Fick's second law in Eq. 4.16.

The solution in Eq. 4.16 to Fick's second law for cases where the thickness is limited can, of course, also be used to derive the regression parameters D_a and C_{sa}. This is done by curve-fitting measured chloride profiles to the curves in Figure 4.8.

4.4.3 *Error function complement (erfc) model with time-dependent* D_a *and constant* C_{sa}

This section includes the definition of average diffusivity D and the age exponent $\alpha \neq n$; Note that D_{aex} is valid only if C_{sa} = constant.

In most currently used empirical chloride-ingress models the diffusivity is treated as time dependent. The time dependence is very much different for different types of diffusion coefficients.

At the beginning of the 1990s empirical models were developed to include the effect that was obvious in data obtained at different exposure times: a time-dependent D_a. Several researchers started to use expressions for this time dependence, such as

$$D_a(t) = D_a(t_0)\left(\frac{t_0}{t}\right)^a = D_{aex}(t_{ex})\left(\frac{t_{ex}}{t}\right)^a \qquad (4.17)$$

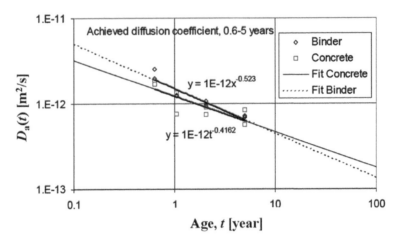

Figure 4.9 'Apparent' or 'achieved' diffusivities plotted against the exposure time. The examples are taken from the chloride profiles in Figure 4.5. The curve-fitting was done in two ways (profiles as % by weight of binder, and profiles as % by weight of concrete), giving two different straight lines for Eq. 4.17.

where t is the age of the concrete, t_{ex} is the age at exposure, $D_a(t_0)$ is the apparent diffusivity at age t_0 and α is the age exponent.

An example of data for 0.6–5 years of exposure is shown in Figure 4.9.

Note that Eq. 4.17 is different from Eq. 4.14, which concerns the instantaneous or point-wise diffusivity in Fick's second law. The similarity, however, causes a great deal of confusion. This will be elaborated further in later sections.

The time dependence was found to be different in different exposure zones. Part of this could be explained by the normal densification of concrete, which was shown laboratory measurements, the densification being different for different binders. However, the phenomenon has not yet been fully explained, and hence one must question whether it is possible to use field data for long-term predictions. An attempt to explain the phenomenon, using time-dependent binding and a physical model, is made in Section 4.5.2.

A time-dependent D causes a great deal of confusion. On one hand, there are observations showing the *apparent* D_a to be time dependent. This value of D is taken from curve-fitting the data after a certain length of exposure to the error function. This means that the instantaneous, or 'point-wise', diffusivity D (D_{F2}) is regarded as constant throughout the whole exposure and that D_a is a kind of an 'average D' ($\bar{D}$) during that period (Figure 4.10).

The time dependence of the constant, apparent $D_a(0, t)$ during an exposure from 0 to t must not be confused with the diffusion coefficient $D(t)$ at

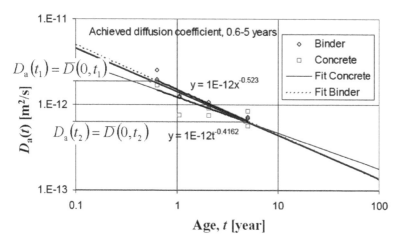

Figure 4.10 The time-dependent apparent D_a after several periods of exposure during which D is assumed to be constant.

a certain age t. $D_a(t)$ is valid for a particular time interval $(0, t)$, but $D(t)$ is valid at a certain age t.

The time dependence of the apparent diffusion coefficient must be explained before it can be used with confidence in predictions. In addition, the age dependence of a D, which is a material property, must be determined, and we need test methods for that. An attempt to clarify these points is made in the next section.

Time dependence of instantaneous diffusion coefficients

A number of measurements show that a chloride diffusion coefficient decreases with time. A reasonable explanation for this is, of course, the continuous cement hydration and binder reactions that densify the concrete over time. An equation commonly used to describe this time dependence of the diffusion coefficient $D(t)$ is (see Eq. 4.14):

$$D(t) = D_{test} \left(\frac{t_{test}}{t} \right)^n \tag{4.18}$$

where t is the age, t_{test} is the age at testing, and D_{test} is the test result at that age. An example is shown in Figure 4.11, with $D_{test} = 10^{-11}$ m²/s at $t_{test} = 30$ days and $n = 0.50$.

Such a time dependence of the diffusion coefficient means that the diffusion coefficient changes continuously with time or, actually, with age. It should be noted, however, that the decrease with time is seen only up to a

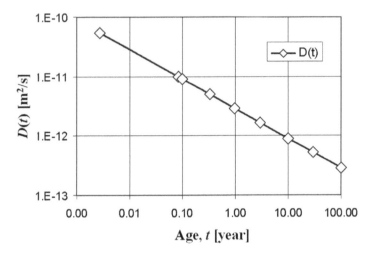

Figure 4.11 An example of the time dependence of the chloride diffusivity. The parameter *t* is the age of the concrete.

limited age, (see Figure 3.23), which may be as little as 6 months or a few years for certain binders.

Time dependence of apparent diffusivities

Numerous field and laboratory data show a clear time dependence of the apparent diffusivity $D_a(t-t_{ex})$. The time dependence is usually expressed in terms of the exposure time $(t - t_{ex})$:

$$D_a(t_{ex}, t) = D_0 \left(\frac{t_0}{t \ \ t_{ex}} \right)^\alpha \qquad (4.19)$$

where *t* is the age, t_{ex} is the age at exposure, D_0 is the diffusivity at a reference age t_0. The exponent α is different from *n* in Eq. 4.18. One obvious peculiarity of this expression is that $D_a(t-t_{ex})$ is not defined at the time of exposure.

It is obvious that the $D_a(t-t_{ex})$ is, and must be, different from $D(t)$, but the relationship between them is not obvious and, amongst other characteristics, it is dependent on the age at exposure. In addition, the continuous densification of the concrete over time does not alone explain the time dependence of the apparent diffusivity.

The apparent diffusivity is time dependent, but, as a consequence of its definition, it is assumed to be constant over the whole exposure time. An example is shown in Figure 4.12, where the time dependence of the instantaneous diffusion coefficient in Figure 4.11 is used to determine a numerical solution to Fick's second law.

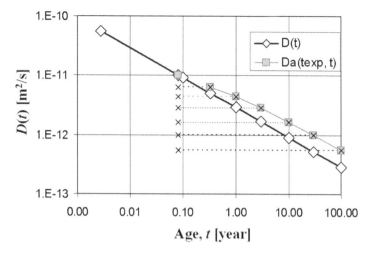

Figure 4.12 An example of the time dependence of the apparent chloride diffusivity. The parameter t is the age of the concrete and t_{ex} is the age at exposure (30 days or 0.082 years).

From Figure 4.12 it is obvious that the time dependence of the apparent diffusivity does not exactly follow an equation like Eq. 4.19. However, such an equation gives an approximate description, although, of course, with a different exponent. An α value of 0.39 gives a fairly good fit for the example in Figure 4.12, as shown in Figures 4.13 and 4.14.

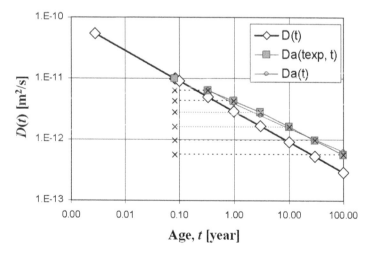

Figure 4.13 An example of an approximate time dependence of the apparent chloride diffusivity according to an equation like Eq. 4.19, for $\alpha = 0.39$.

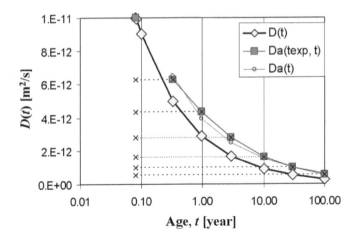

Figure 4.14 The same example as shown in Figure 4.13, but with a linear scale on the vertical axis.

Some researchers have theoretically quantified the relationship between the apparent diffusivity and the instantaneous diffusivity. Visser *et al.* (2002), for example, pointed out the theoretical relationship if the two diffusivities are described by Eqs 4.18 and 4.19, respectively:

$$D_a(t_{ex}, t) = \frac{D_0}{1-n}\left(\frac{t_0}{t-t_{ex}}\right)^n \tag{4.20}$$

This equation, however, is not quite realistic. It is based on another assumption, i.e. that the diffusivity is not defined at the time of exposure. It is said to be valid only for large values of t.

A correct description of $D_a(t-t_{ex})$ can be derived from Crank (1975). For a time-dependent $D(t)$, the $D_a(t)$ times the exposure time $t-t_{ex}$ in the erfc solution is found by integrating $D(t)$ over time, from the start of exposure:

$$D_a(t-t_{ex}) = \int_{t_{ex}}^{t} D(t')dt' \tag{4.21}$$

Inserting Eq. 4.18 into the above equation, the resulting apparent diffusivity will be

$$D_a(t_{ex}, t) = \frac{1}{t-t_{ex}}\int_{t_{ex}}^{t} D_{test}\left(\frac{t_{test}}{t}\right)^n dt \tag{4.22}$$

and the resulting expression will be

$$D_a\left(t_{ex},t\right)=\frac{D_{test}}{\left(1-n\right)}\left(\frac{t_{test}}{t}\right)^{n}\frac{t-t^{n}t_{ex}^{1-n}}{t-t_{ex}}$$

or

$$D_a\left(t_{ex},t\right)=\frac{D_{test}}{\left(1-n\right)}\left(\frac{t_{test}}{t}\right)^{n}\left[t-\left(\frac{t}{t_{ex}}\right)^{n}t_{ex}\right]\frac{1}{t-t_{ex}} \qquad (4.23)$$

A similar expression was derived by Gulikers (2004). The expression is shown in Figure 4.15 to coincide with the result from curve-fitting the numerical solution shown in Figure 4.12.

Different age exponents

The apparent $D_a(t)$ does not follow a straight line in Figure 4.15, and it differs by a factor of up to 2 compared with $D(t)$. An age exponent α for $D_a(t_{ex}, t)$, the slope of a straight line fitted to the curve, would depend on time, and certainly be different from the age exponent n.

The mathematics concerning Fick's second law for chloride ingress has now been completely clarified, by Frederiksen *et al.*(2008), and this should help avoid future misunderstandings. The relationship between the age exponents in the two time-dependent diffusion coefficients have been derived as (Figure 4.16):

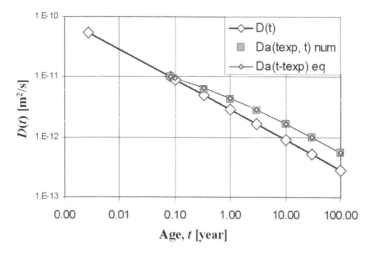

Figure 4.15 An example of a time-dependent apparent chloride diffusivity according to Eq. 4.23, compared with the numerical solution and curve-fitting.

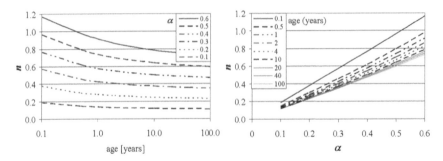

Figure 4.16 The relationship between the age exponents n and α in the two time-dependent diffusion coefficients $D(t)$ and $D_a(t_{ex}, t)$, as a function of age, from Eq. 4.24.

$$n = f(\alpha, t) = \alpha + \frac{\ln\left[(1-\alpha) + \alpha \dfrac{t_{ex}}{t}\right]}{\ln\left(\dfrac{t_{ex}}{t}\right)} \qquad (4.24)$$

The conclusion is that the age exponents n (of the 'instantaneous diffusivity' $D(t)$) and α (of the 'average diffusivity' $D_a(t)$) are quite different, and must never be confused.

Examples of error function complement (erfc) models with time-dependent D_a and constant C_{sa}

Numerous examples can be found in literature of analytical erfc models where the average or achieved diffusivity D_a is assumed to be time dependent, with expressions for $D_a(t)$ like Eq. 4.17, and the achieved surface chloride content C_{sa} is assumed to be constant over time. Early examples of such models are those described by Takewaka and Matsumoto (1988), Maage *et al.* (1995), Mejlbro (1996) and Poulsen (1996). All these models are based on the empirical use of field data to quantify the age exponent α and the diffusivity D_{aex} at an early age (see Eq. 4.17).

DuraCrete (Alisa *et al.*, 1998) used a somewhat different approach. The DuraCrete model for chloride ingress uses a time-dependent average diffusion coefficient $D_a(t_{ex}, t)$, which is the same as in the left-hand part of Eq. 4.17, with a D_0 at an age of t_0.

As the age t_0 of testing is different from the age t_{ex} at exposure, this is another way of describing $D_a(t_{ex}, t)$. The age at exposure t_{ex} does not even appear in the model equation, even though the parameters are evaluated from exposure tests with a certain age at exposure, usually different from

28 days. Obviously, the mathematical expression of this alternative way of describing the average diffusion coefficient is:

$$D_0 = D_{\text{aex}} \left(\frac{t_{\text{ex}}}{t_0} \right)^{\alpha} \qquad (4.25)$$

This gives a simple relationship between the D_{aex} derived from exposure tests and the D_0 determined from a rapid chloride migration (RCM) test. This relationship is not used in the DuraCrete model, but is instead included in the environmental factor. However, as the age at exposure is different in the exposure data, this will add to the uncertainties in the parameters.

Usually, D_0 is quantified by measuring a D_{RCM} in 24 hours in a RCM test at an age of $t_0 = 28$ days. (Note: Because the D_{RCM} is used in the DuraCrete model, Nilsson (2006) misunderstood the principle of the model, and said, wrongly, that it is using the instantaneous diffusivity $D(t)$.) Then D_0 is found by correcting D_{RCM} with an 'environmental factor' k_e, i.e.

$$D_0 = k_e D_{\text{RCM}} \qquad (4.26)$$

In theory, the environmental factor k_e must be quantified from exposure data by using the simple relationship given by Eqs 4.25 and 4.26. This quantification requires not only a huge set of exposure data for different concretes and environments, but also measurements of D_{RCM} at an early age for the same concretes. Such data are seldom available for long-term exposure.

Later, the DuraCrete erfc model was developed into a *fib* model (*fib*, 2000), based on work by Gehlen (2000). Now the environmental factor depends only on temperature, and the relationships given by Eqs 4.25 and 4.26 are abandoned. This must have introduced further errors and uncertainties into the model.

Errors due to neglecting the time dependence of the surface chloride content

Error function complement (erfc) models with time-dependent D_a and constant C_{sa} neglect the time dependence of the surface chloride content. This time dependence is clearly visible in accurate field data, and must be regarded as being supported by observation. Consequently, it should be included in empirical models. If it is not included, errors will arise in the predictions made using erfc models.

These errors have been analysed and quantified by Frederiksen *et al.* (2008). Two main conclusions can be drawn from this analysis. The first concerns the age exponent α. The age exponent in erfc models with constant C_{sa} will be correctly quantified by using expressions like Eq. 4.17, for evaluating

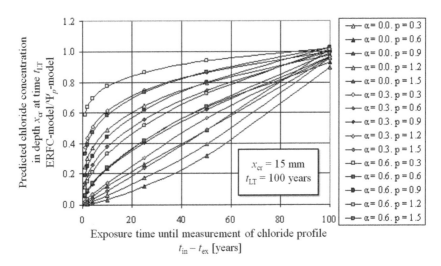

Figure 4.17 The relationship between the predicted and correct chloride content for an erfc model with constant C_{sa} depending on the age exponent α, the time-dependence parameter p for $C_s(t)$, and the last exposure time used. The examples given are for a cover of 45 mm and 100 years service life (Frederiksen *et al.*, 2008).

$D_a(t)$ from field data at different exposure times, even though the erfc models do not correctly describe the time dependence of the C_s values. The second conclusion is much more serious. When applying these models, such as Eq. 4.17, the quantification of the diffusivity (e.g. D_{aex}) at an early age will be significantly erroneous. Examples are shown in Figure 4.17.

The magnitude of this error will depend on the time dependence of the surface chloride content, expressed by the parameter p (see Section 4.4.4), and the age at the latest exposure in the field data. With data for only a few years, a prediction of the diffusivity $D_a(t)$ for 100 years and, consequently, the estimated service life, may be seriously wrong. In realistic cases where the value of p is 0.4–1.1, the error on the derived diffusivity will be a factor 1.5–2 (Frederiksen *et al.*, 2008). This means that, in the past, a significant error was often introduced in the diffusivity determined from exposure data. Therefore, in, for example, the DuraCrete model, the environmental factor is quantified erroneously. Consequently, service-life predictions in such cases may be too optimistic and overestimated by up to two times.

Conclusions on models with time-dependent diffusivities and constant surface chloride content

The time dependence of diffusivities is certainly a possible source of misunderstanding and even mathematical errors. In any model where such a time dependence is used, the time dependence must be very well defined. The age exponents are different for the instantaneous and the average diffusivity, and must not be confused.

Here, only the continuous densification of concrete was taken as a cause of the time dependence of the diffusion coefficients. However, other reasons are possible, and likely, especially under field conditions. These reasons, however, have not yet been fully explained, although an attempt has been made using a physical model and time-dependent binding (see Section 4.5.3).

The assumption of a constant C_{sa} in the erfc models will produce large prediction errors, because the surface chloride content is, in reality, time dependent. As a consequence, empirical and physical models must be able to simulate this time dependence.

4.4.4 Ψ *model with time-dependent* D_a *and* C_s

The erfc model is valid only as long as the surface chloride content C_{sa} is constant with time. Recent findings imply that this assumption must be abandoned, and, as a consequence, another type of model must be used.

Time-dependent surface chloride contents $C_s(t)$

In the 1990s, the apparent surface chloride content C_{sa} was regarded as time dependent by a number of researchers (e.g. Uji *et al.*, 1990; Poulsen, 1996), but at that time no explanation for this was given, but in some environments it is obvious that the chloride content *could* increase with time (e.g. the splash zone in road environments). Drying and wetting in the splash zone of marine structures could possibly explain a time-dependent C_{sa} in that zone.

However, it was observed that $C_{sa}(t)$ is also time dependent in the *submerged* zone of marine structures (see Figure 4.5). The effect is not visible if the data are expressed as chloride by weight of sample, simply because of the large scatter (Figure 4.18), but the time dependence is clearly visible if C_{sa} is expressed by weight of binder (Figure 4.19). This disparity cannot be explained. An understanding of the underlying reasons for the time dependence is vital, both for empirical models and for physical models.

Later exposure data for up to 10 years of exposure have confirmed the time dependence of the surface chloride content for some 40 different concretes (Tang, 2003a,b).

Mejlbro–Poulsen and the HETEK models

The consequence of a time-dependent surface chloride content $C_s(t)$ (see Figure 4.5) is that the error function is no longer the correct mathematical solution to Fick's second law. Instead, a completely different mathematics

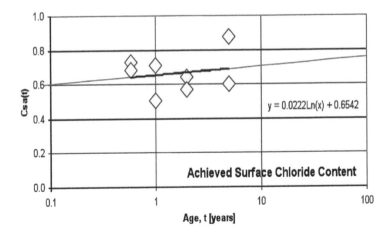

Figure 4.18 Surface chloride contents C$_{sa}$ as a function of exposure times of 0.6–5 years for the same concrete as in Figure 4.5. C$_{sa}$ is expressed by weight of sample.

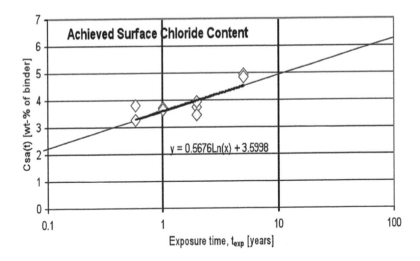

Figure 4.19 Surface chloride contents C$_{sa}$ as a function of exposure times of 0.6–5 years for the same concrete as in Figure 4.5. C$_{sa}$ is expressed by weight of binder.

has to be used. Mejlbro (1996) derived the correct solution and called it the Ψ_p function. This function simply replaces the erfc function in the chloride-ingress equation:

$$C(x, t) = C_s(t_{ex}, t)\Psi_p\left(\frac{x}{2\sqrt{[D_a(t)]t}}\right) \tag{4.27}$$

where the time-dependent diffusivity $D_a(t)$ is the same as in Eq. 4.17. This model was mainly used by Poulsen (1996) and Frederiksen *et al.*(1997a, 2008).

The time-dependent surface chloride content $C_s(t)$ is defined by a parameter p:

$$C_s(t_{ex}, t) = C_i + S\times\left[(t - t_{ex})\times D_{aex}\times\left(\frac{t_{ex}}{t}\right)^\alpha\right]^p \tag{4.28}$$

The exponent p is decisive for how quickly $C_s(t_{ex}, t)$ increases with time, i.e. it depends mainly on the type of binder and the environment. $C_s(t)$ for different values of p is shown in Figure 4.20.

Examples of Mejlbro's solutions to Fick's second law, with a time-dependent diffusion coefficient and time-dependent surface chloride content, for different values of p are shown in Figure 4.21. When $p = 0$ (which is equivalent to C_s being constant), the function is identical to the erfc function.

The parameters t_{ex}, D_{aex}, α and p are required to make a prediction of chloride ingress. In order to make the input data physically understandable, they can be in the form of estimates of C_1, C_{100}, D_1 and D_{100}.

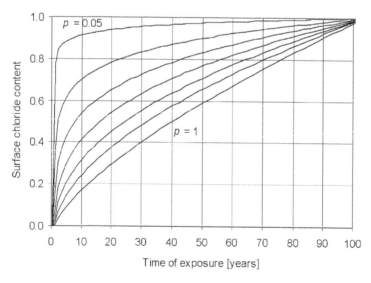

Figure 4.20 An example of the time dependence of $C_s(t_{ex}, t)$ according to Eq. 4.28 (Frederiksen *et al.*, 1997a).

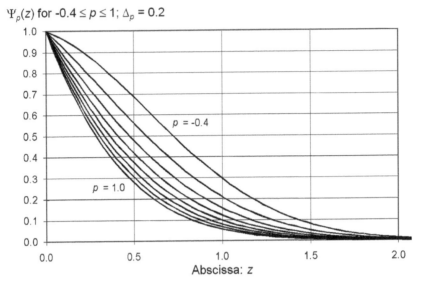

$\Psi_p(z)$ for $-0.4 \leq p \leq 1$; $\Delta_p = 0.2$

Figure 4.21 Graphs of typical Mejlbro–Poulsen Ψ_p functions (Frederiksen *et al.*, 1997a).

Field observations must be used to quantify the time dependence of the surface chloride content $C_s(t)$. For concretes and environments similar to those already studied, excellent models are available. A set of simple models is derived to translate environmental effects into efficiency factors that influence the surface chloride content and the diffusivity (Frederiksen *et al.*, 1997a), taking into consideration the concrete composition and the exposure conditions.

As the parameters are quantified from available field data from one exposure site, the results obtained using the model coincide with real behaviour, limited to that site. However, the parameters can easily be updated if good field data are available from other environments.

One important advantage of this model is that it includes the true solution to Fick's second law, with a time-dependent diffusion coefficient and a time-dependent surface chloride content. Another advantage of the model is that it can be used to estimate all the parameters by employing the concrete composition and the environmental conditions.

4.4.5 *Error function complement (erfc) model with time-dependent* D_a *and* C_{sa}

The 'false erfc model' (Nilsson, 2000) is the simplest empirical model that considers the time dependence of the surface chloride content. The model is, however, purely empirical, and is not a correct mathematical solution to Fick's second law.

The principle of this model is first to estimate the time dependence of the 'apparent diffusion coefficient' $D_a(t)$ and the time-dependent 'apparent surface chloride content' $C_{sa}(t)$ using field exposure tests. The chloride ingress after a certain exposure time t is then predicted using the erfc solution to Fick's second law, with the constant parameters $D_a(t)$ and $C_{sa}(t)$:

$$C(x, t) = C_{sa}(t) \, \mathrm{erfc}\left(\frac{x}{2\sqrt{[D_a(t)]t}} \right) \qquad (4.29)$$

The time-dependent diffusivity $D_a(t)$ is given by Eq. 4.17. The time-dependent surface chloride content is described by an empirical equation including the logarithm of time:

$$C_{sa}(t) = A\ln(\Delta t_{ini} + t - t_{ex}) + B \qquad (4.30)$$

where t is the age, t_{ex} is the age at exposure, Δt_{ini} is an initial binding period (14/365 years), and A and B are regression parameters.

Environmental effects are translated into a surface chloride content $C_{sa}(t)$ mainly by means of exposure tests done in the true environment and for the concrete that is going to be used. Models for time-dependent surface chloride contents are available to some extent, see Section 4.4.7.

The model is very simple and straightforward, the input parameters being quantified from field exposure data obtained under the true conditions. The model parameters can also be updated in a simple way when more field data become available for longer exposure times. It is a true 'empirical' model, and absolutely correct within the limits of the available exposure data. One strong limitation, however, is the lack of mathematical correctness – Eq. 4.29 is not the correct solution to Fick's second law with the boundary conditions given in Eq. 4.30. This leads to an underestimation of chloride ingress, which will be significant if the available data are limited to a few years.

4.4.6 Numerical models with time-dependent D and C$_s$

Solutions to Fick's second law with any time dependence of the diffusion coefficients and the surface chloride content can easily be found numerically. The most elaborated model is Life-365, proposed by Bentz and Thomas (1999). Petre-Lazar *et al.*(2000) have developed a similar model, called the LEO model.

In these models, Fick's second law

$$\frac{\partial C}{\partial t} = D_{F2} \frac{\partial^2 C}{\partial x^2} = \frac{\partial}{\partial x} D_{F2} \frac{\partial C}{\partial x} = -\frac{\partial}{\partial x} J(x) \qquad (4.31)$$

is used as a pure mass balance equation. The instantaneous diffusivity $D(t) = D_{F2}(t)$ is used to describe the flux of chloride ions. This means that the total chloride content is actually the driving potential. This is, of course, physically incomprehensible, but if the chloride-binding capacity is assumed constant (see Eq. 4.9), the model may be used in a correct way.

The diffusivity can be chosen arbitrarily. Life-365 describes the diffusivity as time and temperature dependent:

$$D_{F2}(t, T) = D_{ref}\left(\frac{t_{ref}}{t}\right)^m \exp\left[\frac{U}{R}\left(\frac{1}{T_{ref}} - \frac{1}{T}\right)\right] \qquad (4.32)$$

The boundary conditions are described as $C_s(t)$ in Life-365, increasing at a constant rate up to a certain maximum level after a defined time. However, in principle, any other description could be used. The time dependence according to Eq. 4.32 is used in Life-365 only for a limited exposure time, up to about 25 years. After that, the diffusivity is treated as a constant. Solutions to the mass balance equation with these boundary conditions are found numerically using finite-difference methods.

Part of the time dependence of the diffusion coefficient may be derived from a laboratory test. The environmental effect, however, must be found from field observations, which are used to quantify the time dependence of the surface chloride content $C_s(t)$. For concretes and environments close to the ones already studied, reasonable estimates should be possible. In Life-365 the C_s values are independent of the concrete composition.

There are uncertainties in the definitions and in the quantification of the time dependence of the diffusion coefficient. The exponent m in Eq. 4.32 is usually quantified from observations of the time dependence of the apparent diffusivity, but in other cases it is calculated from measurements on old concretes not exposed to chloride. In Eqs 4.31 and 4.32 the diffusivity is *not* the apparent one. Consequently, the age exponent is different from α (see Section 4.4.3).

4.4.7 Boundary conditions in models based on Fick's second law

In empirical models, the boundary condition is expressed as a surface chloride content C_s, which is really the response by a particular concrete to particular environmental effects. That response depends not only on the environment but also on the concrete. Consequently, an individual C_s value cannot be taken as the boundary condition for a particular environment, as the effect of the concrete must be considered as well (Figure 4.22).

The parameter C_s is the total amount of chloride in the near-surface region of the concrete. Consequently, it depends on the porosity of the concrete and the chloride-binding capacity of the concrete. This explains why C_s is a function of the concrete mix, especially the type of binder, the binder content and

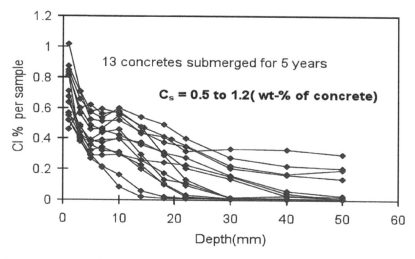

Figure 4.22 Examples of C_{sa} as a function of the concrete mix in one marine, submerged environment (13 concrete mixes submerged for 5 years). (Based on data from Tang (2003b).)

the water/binder ratio. It also explains why C_s is a function of temperature, pH and carbonation.

As the apparent diffusion coefficient obviously depends on the binding capacity (see Eq. 4.12), there is a correlation between D_a and C_{sa} for a particular concrete. Therefore, these parameters cannot be chosen independently.

Lindvall (2003) has developed models for environmental effects, based on the response of particular concretes in particular environments. He used slightly different approaches for marine and road environments.

Models for environmental effects in marine conditions

The surface chloride content in a marine environment is found from an equation such as

$$C_{sa} = C_{sa, eq} k_{C, conc} k_{C, e} \tag{4.33}$$

where $C_{sa, eq}$ is a surface chloride content (% by weight of binder) for an equivalent concrete in submerged conditions in seawater having a chloride concentration of 20 g/l and a seawater temperature of +20°C. Lindvall (2003) found that $C_{sa, eq} = 1.7$ % for a SRPC concrete with a water/cement ratio of 0.40 after 1 year of exposure.

The factors $k_{C, conc}$ and $k_{C, e}$ are correction factors that depend on the concrete composition and the environment, respectively. The concrete factor $k_{C, conc}$ is simply obtained, from data by Frederiksen *et al.*(1997a), as:

$$k_{C, \text{conc}} = 2.5 \left(\frac{w}{C} \right) \tag{4.34}$$

The environmental factor $k_{C, e}$ for submerged conditions was evaluated from a large exposure programme involving concrete specimens submerged in seawater in a number of exposure sites. The surface chloride content was found to depend more on the seawater temperature than on the salinity (Figure 4.23).

The effect of the seawater temperature T was found to be well described by a correction factor $k_{C, T}$, which is expressed by an equation similar to the one for the temperature effect on chloride binding:

$$k_{C, T} = \exp \left[3200 \left(\frac{1}{273 + T} + \frac{1}{293} \right) \right] \tag{4.35}$$

The variation in $k_{C, T}$ is shown by the dashed curves in Figure 4.23.

The effect of the salinity, or the chloride concentration, of the seawater is described by a correction factor $k_{C, Cl}$ according to Figure 4.24. This is, of course, similar to a chloride-binding isotherm.

For tidal, splash and atmospheric zones, Lindvall (2003) introduced correction factors for the height above sea level and the distance to the shore.

For road environments, Lindvall (2003) gives an approach that includes

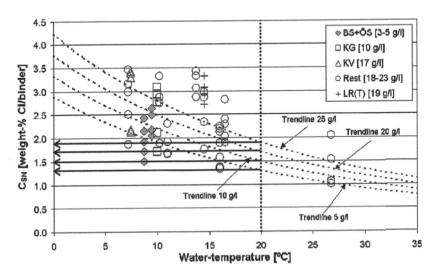

Figure 4.23 The surface chloride content as a function of salinity and seawater temperature, evaluated from a marine exposure programme (Lindvall, 2003).

a number of correction factors as in Eq. 4.33, where $C_{sa,\,eq}$ is defined as the surface chloride content at the road surface next to the outer lane. Correction factors are then applied for the height above the road surface, the distance to the outer lane, traffic speed and intensity, and surface orientation against the morning traffic. An example is shown in Figure 4.25.

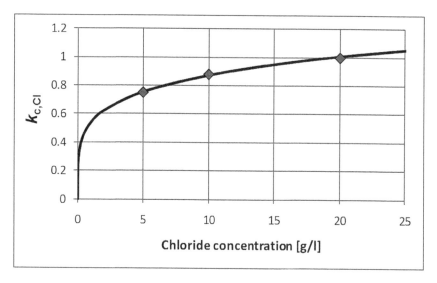

Figure 4.24 The correction factor $k_{C,\,Cl}$ for the chloride concentration of seawater. (Data from Lindvall (2003).)

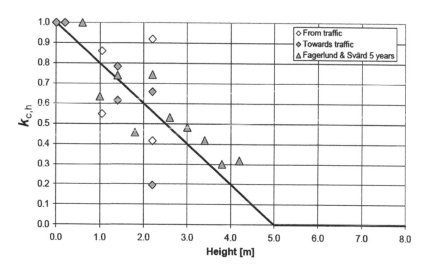

Figure 4.25 The correction factor $k_{C,\,h}$ for the height above the road surface. (Data from Lindvall (2003).)

4.4.8 *Conclusions on models based on Fick's second law*

The discussion in this section has focused on the mathematics of models used to describe chloride ingress into concrete. In practice, high-quality exposure data for long exposure times are urgently needed. Without such data, these models will remain inaccurate, especially with regard to the extrapolation of the age exponents to long exposure times. However, a major improvement would be achieved if it were possible to explain clearly the time dependencies of the diffusion coefficients and the surface chloride contents (see Figure 4.5). To date, such explanations are lacking, and for this reason models based on physics and chemistry are needed so that it is not necessary to depend on empirical comparisons with exposure data. This type of model has been developed during the last 10–15 years.

4.5 Chloride-ingress models based on flux equations

4.5.1 *General*

In 'physical models' or, more strictly speaking, 'mechanistic models', all physical and chemical/electrochemical processes are described as scientifically correctly as possible. True physical models use independently determined input data and no curve-fitting to exposure data. Instead, field exposure data are used to validate predictions. If the comparison between the predicted results and the exposure data is not sufficiently good, the model must be improved, or better data must be determined.

Sophisticated physical prediction models for chloride ingress into concrete contain at least two mass balances and several relationships for chloride and water (Figure 4.26). Among these relationships, the decisive concepts in

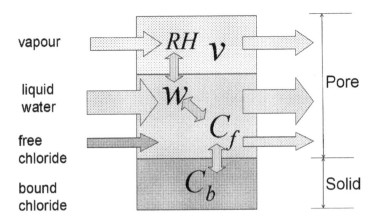

Figure 4.26 The mass balances and interactions of chloride ions and water, in different phases.

physical models are the phenomena associated with chloride transport and chloride binding.

Several input parameters are required for describing chloride transport in physical models. The diffusion coefficients should be described at least as functions of temperature, moisture content, degree of hydration and depth from the surface. The description of the convection term requires the liquid water flow to be expressed adequately.

4.5.2 Boundary conditions in physical models

The description of boundary conditions in physical models is much more complicated than in other models, because physical models require boundary conditions defined in terms of the chloride, temperature and humidity conditions at the surface of the concrete: $C(x=0,t)$, $W(x=0,t)$, $T(x=0,t)$. For the submerged zone of marine structures this is fairly simple, but for all other environments it is extremely complicated. The surface conditions at a particular concrete surface must be regarded as dependent on the location, orientation, distance and time.

There are few data available for the same concrete that has been exposed at two locations. Individual researchers have been exposing test concretes in locations close to their own laboratories, and it is not possible to compare the data collected at these different sites. In addition, there is a lack of proper documentation of the exposure environments. For example, what is defined as a 'splash zone' in one study is not the same as a 'splash zone' in another study.

A possible approach to collecting information on boundary conditions for future use in physical models is to describe the environmental effects in a stepwise fashion, from the meteorological conditions to a description of the surface:

- regional climate
- local climate
- location of the structure
- distance from the source of chloride to the concrete surface
- orientation of the surface.

This approach is illustrated for environmental effects on a road in Figure 4.27. On the regional scale, the macro-climatic effects are those that are not associated with the road at all. On the meso-scale, the environmental effects are those associated with the road surface, but not the actual concrete structure. It is at the micro-scale that the effects of the location, and the size and the shape of the concrete structure are considered, to give the actual environmental effects at a particular concrete surface.

Due to the lack of availability of proper boundary conditions, it is necessary to limit the use of physical models to the response of concrete in only

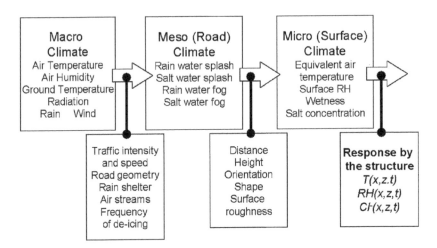

Figure 4.27 An approach to the stepwise quantification of the environmental effects at a concrete surface in a road environment, starting from meteorological data.

some environments and under some conditions. Some data providing a comparison between marine and road climates are presented in Figure 4.28. The build-up of chloride profiles during one season is shown in Figure 4.29 for the same concrete as in Figure 4.28. These data clearly illustrate the challenges of using physical models to describe chloride transport and binding in concrete.

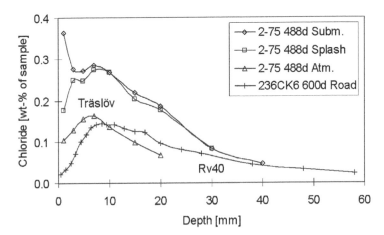

Figure 4.28 Chloride profiles for a SRPC concrete with a water/binder ratio of 0.75 exposed to various chloride environments: a marine submerged (upper curves), and atmospheric and road environment (lower curves).

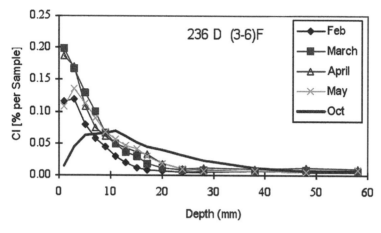

Figure 4.29 Chloride build-up and washout during the first winter and summer of the vertical surface of a concrete with a water/cement ration of 0.75 in a road environment.

In the future, processes similar to those illustrated in Figures 4.28 and 4.29 must be described in order to use physical prediction models for complicated environments. For the time being, these difficulties favour the use of empirical models, because of the simple description of the boundary conditions, i.e. a C_s value. The C_s for each individual concrete could be translated into an 'equivalent' chloride concentration at the concrete surface. This could then be used as the boundary condition for a new concrete, without having to undertake an exposure programme.

4.5.3 Models based on Fick's first law, without convection

The simplest way to create a physical model is, of course, to use Fick's first law (Eq. 4.3) to describe the diffusion of chloride ions. The mass balance requires a description of the interaction between the free chloride ions and the chloride ions bound by the matrix. This can be done in many different ways.

One of the most advanced models for predicting chloride penetration into concrete, ClinConc, was presented by Tang and Nilsson (1994) and Tang (1996b). By using a finite-difference method numerical approach, most of the factors involved in chloride penetration are considered in this model in a relevant and scientific way.

The chloride flux is described by the chloride diffusion coefficient in Fick's first law D_{F1}, calculated from a separate migration test at a certain concrete age. The effect of the densification of the concrete on the diffusion coefficient is included up to a certain age, e.g. 6 months for Portland cement concrete.

The chloride binding is described by a non-linear chloride-binding iso-therm that is a function of pH (see Figure 2.4) and temperature:

$$C_{i,\,j}(\text{bound}) = \exp\left[\alpha_{OH}\left(1 - \frac{[OH]_{i,\,j-1}}{[OH]_{\text{initial},\,j-1}}\right)\right]$$

$$\exp\left[\frac{E_b}{R}\left(\frac{1}{T_j} - \frac{1}{T_0}\right)\right]\left(W_{i,\,j(\text{gel})}\frac{f_b}{1000}c_{i,\,j}^B\right)$$

(4.36)

The leaching of alkalis is described as a pure diffusion of hydroxides out of the concrete, and the pH profile is part of the prediction at every time-step. Only alkaline hydroxides are dealt with (i.e. the pH remains greater than 12.6), and this corresponds to a saturated limewater with a hydroxide concentration $[OH] = 0.043$ mol/l.

The mass balance equations for chlorides and hydroxides are solved by using separate terms for chloride diffusion and chloride binding. At every time step the flux of chloride ions is calculated at each depth using Fick's first law. Using the differences in the flux, the change in the total chloride content is calculated at each depth. By using the binding isotherm, the change in the total chloride content is divided into free and bound chloride. The chloride ingress is shown as profiles of the free and total content of chloride ions.

For a new Portland cement concrete, only one parameter has to be deter-mined, i.e. the diffusion coefficient for the particular concrete at a certain age. The other parameters in the model follow from the mix composition and the exposure conditions. For a new binder, the parameters of the bind-ing isotherm must be given.

Temperature effects

The effect of temperature on binding has an important implication for the predicted chloride profiles. As seawater temperature varies during a year, the chloride profiles will be affected (Figure 4.30). Note that the free chlo-ride concentration is higher at some depth from the concrete surface than at the surface during the summer (at 5.25 years). This is due to the temperature effect on chloride binding, and could be the explanation those observations where a higher chloride concentration is found inside concrete than in the surrounding seawater.

Time-dependent binding

Numerous comparisons have been made between laboratory and field data. The chloride profiles for a number of concretes submerged in seawater have

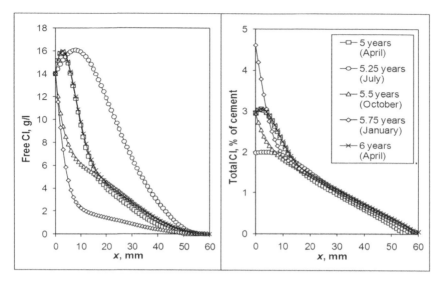

Figure 4.30 Predicted chloride profiles over 5–6 years of exposure, expressed as free and total chloride as determined using the ClinConc model.

been predicted for a certain period of exposure by using the independently determined diffusion coefficients. The predicted profiles coincided well with the field data for up to 2 years of exposure. After this time, for some concretes the measured surface chloride content was greater than the prediction value.

Therefore, the time dependence of the chloride binding had to be empirically included to obtain a better correlation. The chloride binding is assumed to increase with time by a factor of

$$f_t = 1 + a_t \ln(0.5 + t_{Cl}) \qquad (4.37)$$

where t_{Cl} is the duration of the chloride exposure (in years) and a_t is a parameter, which for most concretes is 0.36. This time dependence of chloride binding, however, remains to be explained and quantified independently.

The time dependence of the chloride binding can, of course, explain the time dependence of the apparent surface chloride content $C_{sa}(t)$ (see Figures 4.5 and 4.19). However, the consequence of the time-dependent binding on the chloride ingress is, different than its effect on the time dependent $C_{sa}(t)$ in empirical models (see Figure 4.6).

The time dependence of the apparent diffusivity $D_a(t)$ can, in fact, be fully explained by nothing but time-dependent chloride binding. This is demonstrated in the next section using the physical model ClinConc.

Explaining the time dependence of the apparent chloride diffusivity

The following assumptions are made for the treatment in this section:

- The diffusion coefficient in Fick's first law is constant with time over 100 years, except for the first 6 months.
- Chloride binding is described by a non-linear binding isotherm that is pH dependent (as in Eq. 4.36).
- The time dependence of the chloride binding is described by Eq. 4.37.
- The concrete is submerged with a constant chloride concentration in the surrounding seawater.

Chloride profiles predicted using these assumptions and data for an SRPC concrete (mix code 3–40) with 5% silica fume and a water/binder ratio of 0.40, are shown in Figure 4.31.

The time-dependent chloride binding, of course, causes the surface chloride content C_s to increase with time. By curve-fitting the profiles in Figure 4.31 to the erfc solution to Fick's second law, the regression parameters C_{sa} and D_a can be quantified as a function of time. The function $C_{sa}(t)$ is shown in Figure 4.32.

The achieved diffusivity D_a was also obtained from this curve-fitting (Figure 4.33). Consequently, the time dependence of the chloride diffusivity in empirical models can be explained solely by the time dependence of the chloride binding.

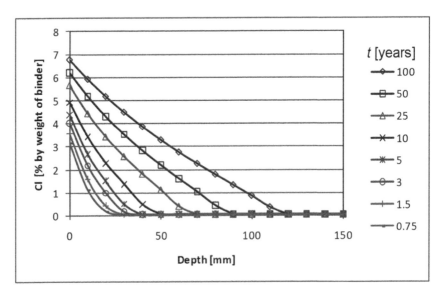

Figure 4.31 Predicted chloride profiles over 100 years for concrete 3–40 with a constant D_{F1} and time-dependent, non-linear and pH-dependent chloride binding.

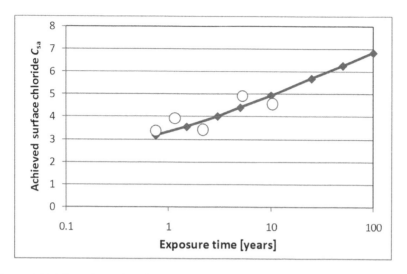

Figure 4.32 The achieved surface chloride content $C_{sa}(t)$ obtained by curve-fitting the predicted profiles shown in Figure 4.31.

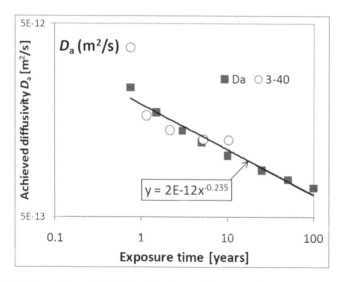

Figure 4.33 The achieved diffusivity D_a obtained by curve-fitting the profiles shown in Figure 4.31. The line is from the traditional exponential relationship (Eq. 4.17).

Advantages and limitations with physical models based on Fick's first law

The most significant advantage of physical models that use Fick's first law as a flux equation, especially the ClinConc model, is the small number of input parameters required, only one of which must be determined and quantified in a separate, independent test. Another advantage is that the effect of temperature variations and leaching can be considered in a physical sense.

However, the physical part of these models can only be used to treat cases of saturated concrete under submerged conditions. The exclusion of the effects of *all* ions on the chloride transport process gives rise to questions about the ability of these models to describe future ingress in a correct way.

4.5.4 Models based on Fick's first law, with convection

A number of similar physical models are available where a convection term has been added (Buenfeld *et al.*, 1995; McLoughlin, 1997; Nilsson, 1997; Denarié *et al.*, 2003; Meijers, 2003; Petre-Lazar *et al.*, 2003). Such models make it possible to predict chloride ingress in a number of situations where moisture flow and wetting and drying play a dominant role. These models involve different simplifications and assumptions, but they are described together here. The examples of details given here are taken from the HETEK convection model (Nilsson 1997, 2000), partly because it is fairly elaborate model, and partly because it is the best known of these models to the author.

The convection of chloride is governed by the moisture flow, and changes in moisture content will change the chloride concentration in the pore water. Consequently, the moisture distribution must be predicted by solving the mass balance equation for moisture. The moisture flow must be separated into two different fluxes, only one of which carries the ions.

The mass balance equation for chloride and the separation into free and bound chloride ions can be described by taking into account the moisture content, because the free chloride ions are limited to the water-filled parts of the pore system. The two mass balance equations are interrelated, and must be solved together. The main part of the solution is to describe the flow of moisture and chloride ions. The initial conditions are trivial, but the boundary conditions must be, until better data are available, a simplification of the true environmental conditions at the concrete surface.

Traditionally, the moisture flow is divided into vapour flow and liquid flow, i.e. with the vapour content and pore-water pressure as driving potentials, respectively:

$$q_w = q_v + q_l = -\delta_v(RH)\frac{\partial v}{\partial x} - k_P(w)\frac{\partial P_w}{\partial x} \tag{4.38}$$

This division of the total moisture flow should not be seen as rigid. Moisture flow in most concretes in chloride environments occurs as a flux of adsorbed water. Part of the adsorbed water may well be able to carry ions, and the division in Eq. 4.38 should be seen as a division into two parts, where only one carries ions, this not necessarily being the flux of only water.

The chloride flux is described by Fick's first law, in the part of the pore system that is filled with water and can carry ions. Thus the diffusion coefficient is a function of the moisture content, and one term giving the convection of chloride within the liquid water flow:

$$q_{Cl} = -D_{Cl}(w, T, c'_f)\frac{\partial c'_f}{\partial x} + q_1 c'_f \tag{4.39}$$

A number of relationships must be expressed mathematically. This is done by considering that the moisture sorption depends on the chloride concentration, and by describing the separation of chlorides into free and bound chloride ions, possibly with a non-linear binding isotherm, including the moisture content in the pore system. All these relationships are expressed differently in the various available models.

The input data required for physical models, at least the most sophisticated ones, are very demanding. The chloride and moisture flow coefficients must be given as functions of numerous parameters. The relationships between these parameters require a large number of input parameters to be quantified. The input data for the boundary conditions require the duration of dry and wet periods at the concrete surface, and the temperature and humidity conditions at the surface during these periods. In addition, the boundary conditions must be described as the concentration of free chloride ions in the solution in contact with the concrete surface, when these are present.

Physical models of this kind are the only models capable of handling cases where moisture flow plays a dominant role. Examples are concrete structures with limited thickness that are exposed to chloride only on one side (e.g. tunnels, pool walls, foundation walls), and structures exposed to drying and wetting, sometimes with splash from de-icing salts (road structures) or seawater (marine splash and atmospheric zones). The models are also capable of handling chloride transport in material combinations. Some examples are given in Figures 4.34 and 4.35.

The predicted behaviour during 'wick action' in Figure 4.34 looks very much like the measurements made by Francy (1998), who could not simulate the peak at a certain depth. The presence and location of this peak in chloride content depends very much on the division of the moisture flow into two parts, where one is carrying ions and the other is not (Nilsson, 2000).

The predictions in Figure 4.35 should be compared with the measurements in Figure 4.29. The measured profiles in specimens along a highway exposed to chloride during the first winter and rain in the first summer look very similar to the predicted profiles in Figure 4.35.

4.5.5 Models based on the Nernst–Planck equation

The development of physical models has continued to include a complete 'multi-species approach', by combining the Nernst–Planck flux equation

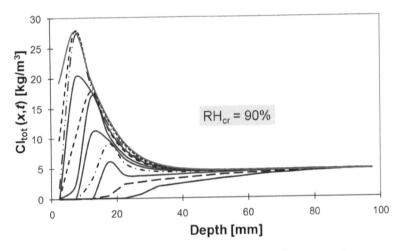

Figure 4.34 Predicted profiles of total chloride obtained from a steady-state experiment with a drying surface (depth = 0) (e.g. wick action). A large portion of the total moisture flow is 'vapour flow. Based on Nilsson (2000).)

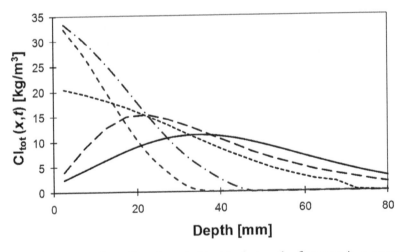

Figure 4.35 Predicted profiles of total chloride during the first year in a concrete structure exposed to de-icing salt. (Based on Nilsson (2000).)

with modules describing the chemical interaction between 'all' the ions in the pore solution and the matrix (e.g. Samson and Marchand, 1999; Johannesson, 2000; Truc, 2000; Marchand *et al.*, 2002; Khitab *et al.*, 2004; Hosokawa *et al.*, 2008).

The fluxes of all the ions in the system are predicted by considering how they influence each other by creating an electrical field in the pore solution.

This electrical field must be predicted simultaneously. The profile of each species is then predicted by solving the mass balance equations for all ions.

The model must describe the flux of every individual type of ion i, which is done using the Nernst–Planck equation (Eq. 4.4). In some models, a convection term is added to take into consideration the effect of liquid moisture flow. Some models even include the effect of degradation and micro-structural changes, which will increase the diffusion coefficients over time.

The flux equations include the electrical field. This electrical field is created by species with charges of opposite sign, so that the faster ions will be slowed down and the slower ones will be speeded up. A general expression for the electrical field (Masi *et al.*, 1997), is given by:

$$\frac{\partial \Phi}{\partial x}(x, t) = -\frac{RT}{F} \frac{\dfrac{I(t)}{SF} + \sum_i z_i D_i \dfrac{\partial c_i(x, t)}{\partial x}}{\sum_i z_i^2 D_i c_i(x, t)} \tag{4.40}$$

where $I(t)$ is the electrical current and S is the cross-section of the material. The Poisson equation may be used instead of Eq. 4.40 in order to determine the electrostatic potential (Marchand *et al.*, 2002):

$$\frac{\partial^2 \Phi}{\partial x^2} + \frac{F}{\varepsilon} \left[\sum_{N} z_i c_i(x, t) \right] = 0 \tag{4.41}$$

where N is the total number of ionic species and ε is the dielectric permittivity of the medium.

For each type of ion, the mass balance equation is solved:

$$p_{\text{sol}} \frac{\partial c_i(x, t)}{\partial t} = -\frac{\partial q_i(x, t)}{\partial x} - \frac{\partial C_{\text{b}, i}(x, t)}{\partial t} \tag{4.42}$$

where the right-hand term gives the binding/interaction between species i and the matrix (e.g. chloride binding or some kind of reaction or liberation).

The input data required for this kind of physical model is very demanding, especially if convection is considered. The ion diffusion coefficients and moisture flow coefficients must be given as functions of a number of parameters. As in the case of Fick's first law with convection (see Section 4.5.4), large number of input parameters must be quantified to establish the relationships between the parameters in the model. All the other challenges highlighted for Fick's first law with convection in Section 4.5.4 are applicable here as well.

In addition, the parameters of the binding/interaction term must be quantified separately. Different models use various ways to consider chloride binding. Truc (2000) used a chloride-binding isotherm, Samson and Marchand (1999) focused on the chemical fixation of chloride, and Johannesson (2000) described the chloride binding as a special ion–solid interaction process. Hosokawa *et al.* (2008) used a thermodynamic database for the equilibrium between dissolved species, vapours and mineral phases.

The diffusion coefficients D_i for each type of ion cannot be determined in separate tests, because the flux of one ion depends on both the presence and the concentrations of the other ions. Samson and Marchand (1999) overcame this obstacle by determining a 'formation factor' for the pore system using another test method, and then simply multiplied the formation factor by the diffusion coefficient for each type of ion in the solution. If this is correct, a test method for the formation factor is required. Khitab *et al.* (2004) proposed measuring the chloride diffusion coefficient using a migration test (LMDC test, developed at the Laboratory of Materials and Durability of Constructions (LMDC), Toulouse, France). The diffusion coefficients of the other ionic species are related to the chloride diffusion coefficient in the same way as in an infinitely dilute solution.

The boundary conditions at the concrete surface must describe the environmental effects as a function of time in terms of the time-varying chloride concentration where there is a solution in contact with the surface, alternating with drying conditions. To date, no models are available for non-trivial cases that are able to translate actual environmental effects into the required boundary conditions at the concrete surface. In practice, this means that only simple boundary conditions can be treated, such as direct contact with a salt solution or drying in an atmosphere of a given humidity.

Physical models of this kind predict the moisture profiles, if convection is considered, and the concentration profiles of the free and total ions. An example of predicted ion profiles, ion fluxes and electrical potential for a 3 cm thick concrete specimen in a chloride migration test is shown in Figure 4.36.

Physical models of this kind are the only models capable of considering the effects of all ions in a correct way. This makes them especially suitable for explaining observations made in laboratory and field tests that cannot be foreseen by simple empirical and physical models. If physical models are used in this way, the confidence in the results obtained using simple models could be significantly improved.

4.5.6 *Conclusions on models based on flux equations*

The main conclusions regarding physical models are:

- Physical models, especially those that consider all ions and convection at the same time, give theoretical predictions as close to the present experimentally obtained knowledge as possible.
- Physical models may be used to check the relevance of more simple, empirical models, and may explain some of the assumptions that have

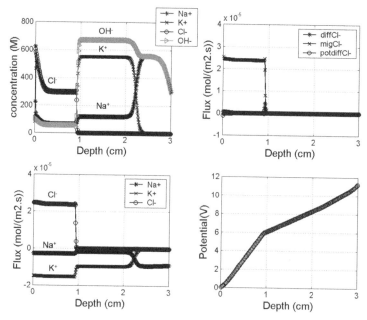

Figure 4.36 Predicted concentration profiles and fluxes for chloride, potassium, sodium and hydroxyl ions in a 3 cm thick specimen, obtained in a migration test where 12 V potential difference was applied. Data are the predicted values when the chloride ions have penetrated the specimen by approximately 1 cm. (Based on Truc (2000).)

to be made in simple models. This may improve the confidence in the results obtained using empirical models.

- Physical models require the quantification of a number of parameters associated with the material tested and its environment, and each of these parameters is a function of several input parameters in the models.
- The required surface climatic conditions are extremely complicated to model.
- As most physical models require a huge amount of input data, which are not readily available at present, and as they have limited user-friendliness, most physical models will remain as research tools and not be used in practical applications.

5 Sensitivity analysis and tests of chloride-ingress models

5.1 Introduction

It has been observed by the authors that, in the literature, chloride-ingress models are normally tested on limited exposure data. As almost all models have parameters that can be adjusted to show good agreement of predicted values with the field data, the limitations of the models or areas where further knowledge is needed are not identified. In this chapter, examples are presented of the independent sensitivity analyses of models, and of comparing prediction results independently with exposure data. The models that have been tested are:

- the error function complement (erfc) model, with a constant diffusion coefficient and a constant surface concentration;
- the LEO model, an empirical model developed by EDF in France in 1998;
- the multi-species diffusion (MsDiff) model.

For more information on these models, see Chapter 4.

5.2 Probabilistic sensitivity analysis: examples

The analysis in this section is largely the same as the one that was done in WP 4.2 of the EU project CHLORTEST (Nilsson, 2005). It was mainly performed by the study partners at INSA in Toulouse and EDF in Paris, details of which can be found in the work package report WP 4 of the EU project CHLORTEST (Nilsson, 2005).

In the next section, the methodology used to determine sensitivity of models for predicting chloride penetration in the short-term (0–10 year exposure) is presented. When carrying out a sensitivity analysis it is essential to:

- define a practical situation
- define the model
- define the input data and the uncertainty in the data (for the practical situation considered)

- define a limit state
- calculate probabilities (run the model with several sets of input data).

This process will lead to the identification of the uncertainty in the final result and the importance of the input data.

5.2.1 Methodology for sensitivity analysis

Usually, the input data for a model are random variables. The environmental parameters are not very well defined and, even if the concrete formulation is well known, its properties vary in all construction projects. However, a deterministic approach is not enough, and it is important to know the influence of the variation in input data on the output data for each model. A probabilistic approach can help to evaluate these variations.

Principle of the probabilistic method

In general, the intention is to search the probability of a function $G(X)$ (where X is the random variable vector of the input data for a model) to find values of this function that are greater than 0. $G(X) = 0$ is the limit state function and separates the input data into two sets: the failure set ($G(X) < 0$) and the safe set ($G(X) > 0$). Different numerical methods exist, known as simulation methods (e.g. the Monte-Carlo method, FORM and SORM, first- and second-order reliability methods) to resolve this kind of problem. These methods are available in the probabilistic software PROBAN (Olesen, 1992), developed by Det Norske Veritas. Using this software it is possible to determine the failure probability ($P(G(X) < 0)$, the most probable failure point and the importance factors (Madsen *et al.*, 1986). Note that, for each set of input data, the importance factor measures the relative influence of the variability in the data on the probability of failure. This means that the higher the importance factor is for a set of input data, the greater is the influence of the data variability on the position within the safe/unsafe set.

REFERENCE TO PROBABILITY LAW OF A RANDOM VARIABLE

Mathematically, a random variable is defined as a value that depends on the result of a random test, i.e. the value cannot be judged in advance with good accuracy. The law of probability of a random variable allows the values to be calculated. A probability law F_X can be characterised by a repartition function R:

$$F_X: R \rightarrow [0,1] \qquad x \rightarrow F_X(x) = P(X \leq x) \tag{5.1}$$

where X is a random variable, x is its realisation and F_X is an increasing function, which is continuous on the right-hand side as

$$\lim_{x \to -\infty} F_X(x) = 0 \text{ and } \lim_{x \to +\infty} F_X(x) = 1$$

Therefore, it is possible to calculate the probability that X will belong to any interval of R:

$$\forall (a, b) \in R^2, a < b, P(a < X \le b) = F_X(b) - F_X(a) \tag{5.2}$$

Random variables are classified as a function of their value validity set. Therefore, a random variable X belongs to a Gaussian distribution law (normal law) if its values are in R and

$$F_X(x) = \frac{1}{\sigma\sqrt{2\pi}} e^{\frac{-(x-m)^2}{2\sigma^2}}$$

The beta distribution law is expressed as follows:

$$F_X(x) = \frac{(x-a)^{r-1}(b-x)^{t-r-1}}{(b-a)^{t-1} B(r, t-r)} \tag{5.3}$$

with

$$B(r, s) = \int_0^1 t^{r-1}(1-t)^{s-1} dt \tag{5.4}$$

This last choice allows the correct modelling of the data, as defined by 'an inferior limit, a most probable value and a superior limit'. The parameters a and b correspond to these limits, and x_{max} is the most probable value in the set $[a, b]$. Therefore, the mean value m and the standard deviation σ of the variable are:

$$m = \frac{a + 4x_{max} + b}{6} \text{ and } \sigma = \frac{b-a}{6} \tag{5.5}$$

Definition of a practical situation

A practical situation has to be defined by one concrete located in one particular environment for a defined service life. The case considered here is of free diffusion of chloride ions in a saturated concrete. The input data for a particular model are based on the experimental data corresponding to this situation obtained for the defined concrete.

The environment is pure water with NaCl, $[Cl^-] = 18$ g/l in the upstream compartment and a zero chloride concentration in the downstream compartment of a diffusion test set-up.

The simulations are made for the concrete defined in Table 5.1.

The composition of the initial solution in the concrete pores is 23 mol/m^3 Na$^+$, 156 mol/m^3 K$^+$, 0 mol/m^3 Cl$^-$ and 179 mol/m^3 OH$^-$.

Table 5.1 Characterisation of the concrete used for the simulation

Description	Unit	Quantity
Cement	kg/m^3	560
Sand	kg/m^3	695
Gravel	kg/m^3	825
Water	Litre/m^3	224
Paste	kg/m^3	784
Water-accessible porosity[1]	%	16
Solid density	kg/m^3	2710
Concrete hydration coefficient		$\alpha_h = \dfrac{(w/c) - p_{sol}\left[(w/c) + 0.32\right]}{0.36} = 0.79$
		[Powers' formula]
Concrete dry density	kg/m^3	2304

Note
1 As determined based on the quantity of water intake under the 'saturated condition'.

The concrete effective diffusion coefficient D_e was measured in 28-day-old concrete, using the INSA test (as D_{ssm} in Chapter 2):

- Sample 1: $D_{e1} = 1.71 \times 10^{-12}$ m^2/s.
- Sample 2: $D_{e2} = 1.72 \times 10^{-12}$ m^2/s.
- Sample 3: $D_{e3} = 1.76 \times 10^{-12}$ m^2/s.

The coefficient D_e is a random variable having a normal distribution. The mean value is:

$$D_{e,mean} = \frac{(1.71 + 1.72 + 1.76) \cdot 10^{-12}}{3} = 1.73 \times 10^{-12} \quad \text{m}^2/\text{s}$$

and its coefficient of variation (COV) is 1.25% (Note: This value is just an example. The actual variation from the inter-laboratory comparisons is much larger (see Table 3.13 in Chapter 3).)

Model definition

A model allows the prediction of the chloride penetration as a function of time, using as input data the concrete properties and information about the environment. The sensitivity of three models was studied:

- the erfc model with constant diffusion coefficient and constant surface concentration
- the LEO model
- the MsDiff model.

For some models, the input data show a wide variability, and a probabilistic approach was used (Petre-Lazar, 2000). Each parameter is defined by its distribution law. For the measured data, a Gaussian distribution is proposed. For the parameters that are not very accurately known, but for which the set of variation is defined, a beta distribution is proposed. In the case considered here, the influence of two particular pieces of input data, which are used in all the models, the chloride diffusion coefficient and the chloride surface content, was studied.

Definition of limit state

In the probabilistic method (FORM) a limit state is required in order to perform the sensitivity analysis. This limit state defines the transition between the safe set and the unsafe set of data. Nevertheless, the choice of the limit state has a great influence on the sensitivity analysis. So, in the case considered here, a limit state regarding steel corrosion is proposed. This limit state is defined as a critical concentration of chloride that initiates corrosion on the steel surface within the concrete (even though the definition of such a value is still under discussion). In addition, limit states of a cover depth of 5 cm from the concrete exposed surface, and a critical chloride concentration (expressed by volume of pore solution) of $c_{crit} = 6.5$ g Cl$^-$/l are chosen. Thus it is considered that corrosion is initiated at a depth of 5 cm when the chloride concentration is greater than $c_{crit} = 6.5$ g/l. The state limit function can then be written as:

$$G = c_{crit} - c(x = 5 \text{ cm}, t) \qquad (5.6)$$

Analysis execution

The Monte-Carlo simulation and the FORM/SORM method are used to undertake the probabilistic analysis. The Monte-Carlo method is based on a large number of random input data sets. For each set, the response of the model is calculated, and then a statistical treatment is applied to the whole of the results. The FORM/SORM and Monte-Carlo methods allow the calculation of the importance factors.

The principle of the FORM/SORM method is illustrated in Figure 5.1. The FORM method is used to:

- evaluate the fiability index β; the design point U^* is determined by an optimisation algorithm:

$$U^* = \arg \min \left\{ \|U\|^2 \,|\, G(U, S(U)) \leq 0 \right\} \qquad (5.7)$$

- obtain an approximate value of the failure probability:

$$P_f \approx \Phi(-\beta) \qquad (5.8)$$

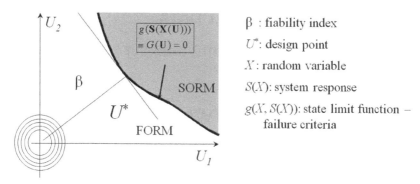

β : fiability index

U^*: design point

X : random variable

$S(X)$: system response

$g(X, S(X))$: state limit function – failure criteria

Figure 5.1 Geometrical representation of the conception point.

To increase the precision of Eq. 5.8, the SORM method has been proposed. The idea is to replace the limit state surface by a quadratic surface for which the probabilistic content is analytically known:

$$P_f \approx \Phi(-\beta) \prod_{i=1}^{n-1} \frac{1}{\sqrt{1 + \beta k_i}} \tag{5.9}$$

The Monte-Carlo method can be represented as in Figure 5.2. These methods are available in the PROBAN software.

Figure 5.3 illustrates the realisation of the sensitivity analysis. The output is a couple of values that characterise the range of the output data. The

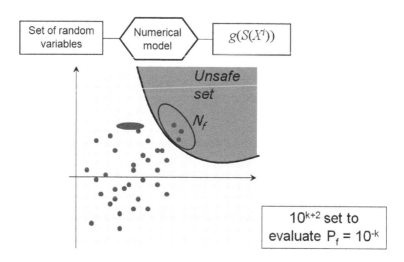

Figure 5.2 Illustration of the Monte-Carlo method.

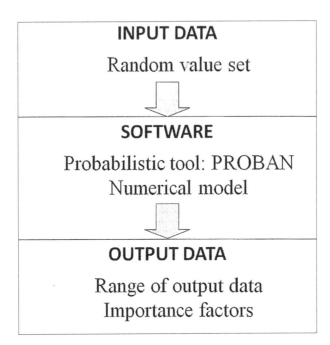

Figure 5.3 Representation of the sensitivity analysis.

fractiles at 95% and 5% are calculated. These are determined by the following relations:

$$f_{95} = m + 1.645\sigma \text{ (superior limit)}$$

and

$$f_5 = m - 1.645\sigma \quad \text{(inferior limit)}$$

where m is the mean value and σ is the standard deviation of all the obtained output data.

5.2.2 Application of the methodology in the error function complement (erfc) model

Definition of the model

The erfc model is based on the straightforward error function:

$$c(x, t) = c_s \left[1 - \text{erf}\left(\frac{x}{2\sqrt{D_a t}} \right) \right] \tag{5.10}$$

where $c(x, t)$ is the chloride concentration over time (g of Cl/l), c_s is the chloride concentration in the external solution (g of Cl/l), D_a is the apparent diffusion coefficient (m²/s), t is the exposure time (s) and x is the penetration depth (m).

Input data

The input data for the ERFC model are given in Table 5.2.

Results for the erfc model

CHLORIDE PROFILES

The simulated chloride profiles obtained using the erfc model are shown in Figure 5.4.

Table 5.2 Input data for the erfc model

No.	Parameter	Meaning	Statistical distribution
1	c_s	Chloride surface concentration	Gaussian distribution Mean value = 18 g Cl/l Coefficient of variation = 5%
2	D_a	Diffusion coefficient	Gaussian distribution Mean value = 4.96×10^{-12} m²/s Coefficient of variation = 1.25%
3	x	Penetration depth	0.05 m (protection thickness)
4	c_{crit}	Limit state of corrosion initiation	6.5 g Cl/l

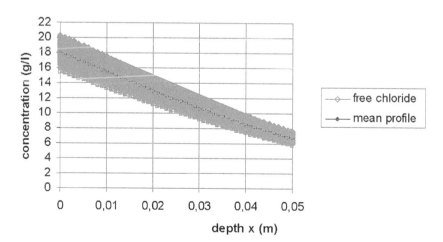

Figure 5.4 A band of chloride profiles at 10 years, obtained using the erfc model.

IMPORTANCE FACTORS

In the case considered, the interest is to determine the importance factors up to the concrete exposure age of 10 years. The importance factors obtained with the erfc model are presented in Table 5.3 and Figure 5.5. It can be seen from Figure 5.5 that the importance factors are time dependent. In the case considered, the variation in the chloride surface concentration is more important than the variation in the chloride coefficient.

CURVE OF THE LIMIT STATE FUNCTION

Figure 5.6 shows the curve of the limit state function that defines the transition between the safe set (passivation zone) and the failure set (non-passivation zone). To obtain more than 10 years of service life, the D value in the erfc model must be smaller than 5×10^{-12} m²/s, when exposed to sea-water with a chloride concentration of 20 g/l.

Table 5.3 The evolution over time of the importance factors for the erfc model

t (years)	Importance factor	
	C_s	D_a
3	69.4	30.6
4	80.0	20.0
5	87.2	12.8
6	91.8	8.2
7	94.5	5.5
8	96.1	3.9
9	97.2	2.8
10	97.8	2.2

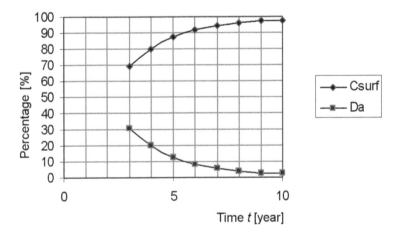

Figure 5.5 The evolution over time of the importance factors for the erfc model.

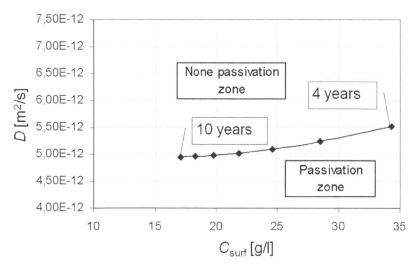

Figure 5.6 Curve of the limit state function for the erfc model.

5.2.3 Application of the methodology in the LEO model

Definition of the model

The function of the model is as follows:

$$c(x, t) = c_{\text{ini}} + (c_{\text{s}} - c_{\text{ini}}) \left[1 - \text{erf} \left(\frac{x}{2\sqrt{\alpha \eta D_{\text{Cl}} t}} \right) \right] \tag{5.11}$$

where c_{ini} is the initial chloride concentration in the pore solution (g Cl/l), D_{Cl} is the effective diffusion of chloride in the pore solution (m²/s), α is the correction coefficient for the interaction of the ionic flux, and η is the correction coefficient for the ion–matrix interaction.

Input data

The input data for the LEO model are given in Table 5.4.

Results of the LEO model

CHLORIDE PROFILES

The simulated chloride profiles obtained using the LEO model are shown in Figure 5.7.

Table 5.4 Input data for the LEO model

No.	Parameter	Meaning	Statistical distribution
1	c_{ini}	Initial chloride concentration in the pore solution	0 (g/l)
2	c_s	Chloride surface concentration	Gaussian distribution Mean value = 18 g/l solution Coefficient of variation = 5%
3	D_{Cl}	Effective diffusion coefficient in the pore solution	Gaussian distribution Mean value = 10.81×10^{-12} m²/s Coefficient of variation = 1.25%
4	x	Penetration depth	0.05 m (protection thickness)
5	c_{crit}	Limit state of corrosion initiation	6.5 g/l

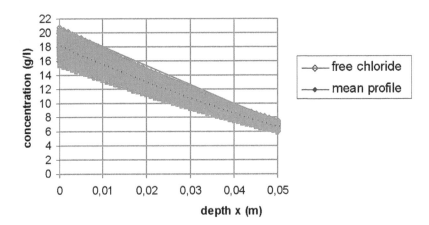

Figure 5.7 A band of chloride profiles at 10 years, obtained using the LEO model.

Importance factors

The importance factors obtained using the LEO model are presented in Table 5.5 and Figure 5.8. As in the case of the erfc model, the importance factors are time dependent and the variation in the chloride surface concentration is more important than the variation in the chloride coefficient.

CURVE OF THE LIMIT STATE FUNCTION

The curve of the limit state function for the LEO model is shown in Figure 5.9. Here, to obtain more than 10 years of service life the D value must be smaller than 1.1×10^{-11} m²/s, when exposed to seawater with a chloride concentration of 20 g/l.

Table 5.5 The evolution *over time* of the importance factors for the LEO model

t (years)	Importance factors	
	c_s	D_{Cl}
4	62.7	37.3
5	72.8	27.2
6	81.5	18.5
7	87.8	12.2
8	92.0	8.0
9	94.6	5.4
10	96.3	3.7

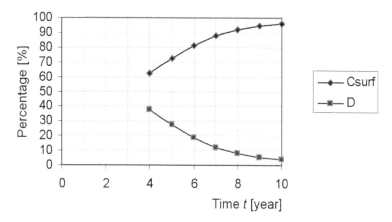

Figure 5.8 The evolution over time of the importance factors for the LEO model.

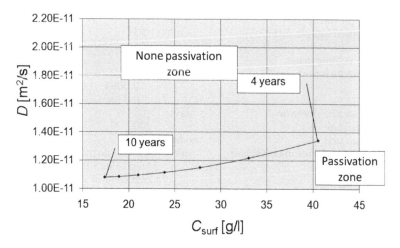

Figure 5.9 Curve of the limit state function for the LEO model.

5.2.4 Application of the methodology in the MsDiff model

Definition of the model

The flux of the ionic species is described by Nernst–Planck equation:

$$\rho\frac{\partial c_i(x,t)}{\partial t} + (1-p_0)\rho\frac{\partial C_{mi,B}(c_i(x,t))}{\partial t} =$$

$$-\frac{\partial}{\partial x}\left[D_{e,i}\frac{\partial c_i(x,t)}{\partial t} + D_{e,i}\frac{c_i(x,t)}{\gamma_i(x,t)}\frac{\partial\gamma_i(x,t)}{\partial x} + z_i D_{e,i}c_i(x,t)\frac{F}{RT}\frac{\partial\varphi(x,t)}{\partial x}\right]$$

(5.12)

where $D_{e,i}$ is the diffusion coefficient of ionic species i (m²/s), c_i is the concentration of ionic species i (mol/m³), γ_i is the activity coefficient of ionic species i, z_i is the valence of ionic species i, φ is the electric potential (V), F is the Faraday constant (= 96 480 J/(V mol)), R is the perfect gas constant (8.314 J/(mol K)), T is the absolute temperature (K), p_0 is the porosity of the concrete, ρ is the density of the concrete when dry (kg/m³) and $C_{mi, B}$ is the number of bound species divided by the mass of dry material (mol/kg).

Four ionic species, Na⁺, K⁺, OH⁻ and Cl⁻, were considered in the MsDiff model.

Input data

The input data for the MsDiff model are listed in Table 5.6.

Table 5.6 Input data for the MsDiff model

No.	Description	Unit	Value, uncertainty
1	Sample thickness	m	0.2
2	Exposed sample surface	cm²	95
3	Porosity	%	16
4	Dry volume mass	kg/m³	2710
5	Freundlich isotherm coefficients		
	α		0.0064
	β		0.33
6	Immersion duration	days	
7	Time step	s	200.000
8	Space step	nodes	100
9	Chloride diffusion coefficient (LMDC test)	m²/s	Gaussian distribution Mean value = 1.73×10^{-12} m²/s Coefficient of variation = 1.25
10	Ratio D_{ion}/D_{Cl^-}		
	Na⁺		0.05
	K⁺		0.0735
	OH⁻		2.73
11	Species concentration in the upstream compartment		

Na$^+$		mol/m^3	Cl$^-$
K$^+$		mol/m^3	0
Cl$^-$		mol/m^3	Gaussian distribution
			Mean value = 18 g/l = 507 mol/m^3
			Coefficient of variation = 5%
OH$^-$		mol/m^3	0

12 Species concentration in the interstitial solution and the downstream compartment

Na$^+$		mol/m^3	23
K$^+$		mol/m^3	156
Cl$^-$		mol/m^3	1
OH$^-$		mol/m^3	178
13	External current	mA	0

Results for the MsDiff model

CHLORIDE PROFILES

The simulated chloride profiles obtained using the MsDiff model are shown in Figure 5.10.

IMPORTANCE FACTORS

The importance factors obtained for the MsDiff model are presented in Table 5.7 and Figure 5.11. As for the erfc and the LEO models, the importance factors are time dependent and the variation in the chloride surface concentration is more important than the variation in the chloride coefficient.

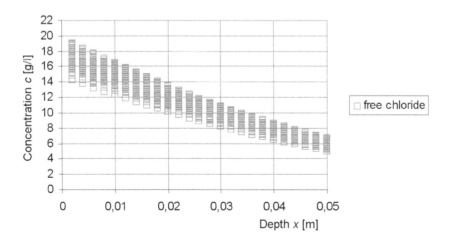

Figure 5.10 A band of chloride profiles at 10 years, obtained using the MsDiff model.

Table 5.7 The evolution over time of the importance factors for the MsDiff model

t (years)	Importance factor	
	c_s	D
4	62.7	37.3
5	72.8	27.2
6	81.5	18.5
7	87.8	12.2
8	92.0	8.0
9	94.6	5.4
10	96.3	3.7

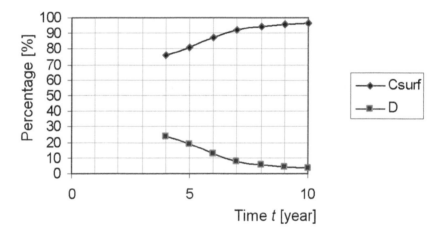

Figure 5.11 The evolution over time of the importance factors for the MsDiff model.

CURVE OF THE LIMIT STATE FUNCTION

The curve of the limit state function for the MsDiff model is shown in Figure 5.12. With this model, to obtain more than 10 years of service life the D value must be smaller than 1.8×10^{-12} m²/s, when exposed to seawater with a chloride concentration of 20 g/l. Again, as in the erfc and LEO models, the importance factors are time dependent and the variation in the chloride surface concentration is more important than the variation in the chloride coefficient.

5.2.5 Conclusions

The sensitivity analysis was undertaken using probabilistic methods and two sets of input data: the surface chloride concentration and the chloride diffusion coefficient. The results of the analysis show that the importance factors

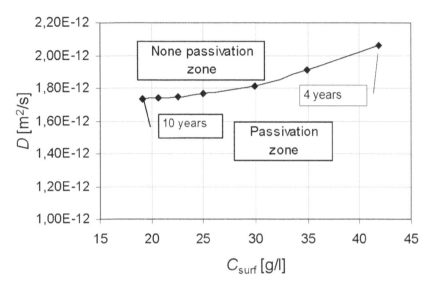

Figure 5.12 Curve of the limit state function for the MsDiff model.

are dependent on time, and this was the case with all three models tested, despite the fact that they are very different conceptually.

In the case considered, and with all three models tested, the influence of the surface concentration of chloride ions is more important than the influence of the diffusion coefficient. This means that it is essential to spend more effort in increasing the precision of methods used to measure the surface concentration of chloride ions in order in order to increase the precision of the prediction. The surface concentration of chloride ions is a function of the environment, but also of the interactions between the cement paste and the chloride ions. Therefore, knowledge of the binding isotherm is key for the prediction of chloride ingress, and is as important as in the prediction of the diffusion coefficient.

The above conclusions are based on the sensitivity analysis done for exposure times of up to 10 years. It is important to remember that the conclusions are dependent on the exposure time, and that for longer exposure times, such as 100 years, the conclusions may be different. This is tested in the next section.

5.3 Long-term sensitivity of error function complement (erfc) models

A second sensitivity analysis is now undertaken in such a way that it covers also much longer exposure times than 0–10 years.

5.3.1 *Mathematical expressions*

As the initial chloride concentration C_i in an erfc model changes only the vertical position of a chloride distribution profile, and not the shape of the profile, in order to simplify the mathematical expressions it is reasonable to assume that $C_i \approx 0$. Thus all the models based on the erfc can be expressed in a general form as follows:

$$C = C_s \text{erfc}(z) \tag{5.13}$$

where C could be further expressed as $(C - C_i)$ and C_s as $(C_s - C_i)$ in the application of the models. In Eq. 5.13,

$$z = \frac{x}{2\sqrt{kD_0 \left(\dfrac{t_0'}{t}\right)^{n'} t}} \tag{5.14}$$

or, more correctly,

$$z = \frac{x}{2\sqrt{\dfrac{kD_0}{1-n}\left(\dfrac{t_0'}{t}\right)^{n}\left[\left(1 + \dfrac{t_{ex}'}{t}\right)^{1-n} - \left(\dfrac{t_{ex}'}{t}\right)^{1-n}\right] t}} \tag{5.15}$$

where t' represents the concrete age and t represents the exposure time. For erfc models with time-dependent $D(t)$ and constant C_{sa}, the combination of Eqs 5.13 and 5.14 is used; for erfc models with time-dependent $D_a(t)$ and constant C_{sa} and those with time-dependent $D_a(t)$ and $C_{sa}(t)$, let $k = 1$ in Eq. 5.14; and for erfc models with constant D_a and C_{sa}, let $k = 1$ and $n = n' = 0$ in either Eq. 5.14 or Eq. 5.15.

Even though in a model with time-dependent $D_a(t)$ and $C_{sa}(t)$, the surface concentration C_s is a function of time t, it will be seen later that this will influence only the magnitude of C_s, and not the sensitivity of C_s for the prediction of C.

5.3.2 *Sensitivity of various parameters in the prediction of chloride concentration*

Surface concentration C_s

The sensitivity coefficient of parameter C_s is:

$$\frac{\partial C}{\partial C_s} = \text{erfc}(z) \tag{5.16}$$

Therefore,

$$\frac{\Delta C}{\Delta C_s} \frac{C_s}{C} = 1 \tag{5.17}$$

This implies that any relative change in surface concentration C_s will result in an equal relative change in concentration C. From Eq. 5.17 it can be seen that the value of $(\Delta C/\Delta C_s) \times (C_s/C)$ has no dependence on any other parameter, such as t. Therefore, regardless of whether C_s is time dependent or not, its sensitivity is the same.

Diffusion coefficient D_0

The sensitivity coefficient of parameter D_0 can be derived as:

$$\frac{\partial C}{\partial D_0} = \frac{C_0}{\sqrt{\pi}} \frac{z}{D_0} e^{-z^2} \tag{5.18}$$

resulting in the following equation:

$$\frac{\Delta C}{\Delta D_0} \frac{D_0}{C} = \frac{1}{\sqrt{\pi}} \frac{C_s}{C} z e^{-z^2} = \frac{1}{\sqrt{\pi}} \frac{z e^{-z^2}}{\frac{C}{C_s}} \tag{5.19}$$

Apparently, the sensitivity of parameter D_0 is related to the ratio C/C_s and the value of z that contains all the variations except for C_s. It can be seen from Eq. 5.19 that $(\Delta C/\Delta D_0) \times (D_0/C)$ is inversely proportional to C/C_s. The quantitative relationships of $(\Delta C/\Delta D_0) \times (D_0/C)$ and z are shown in Figure 5.13. The maximum influence of z on $(\Delta C/\Delta D_0) \times (D_0/C)$ occurs at $z \approx 0.7$.

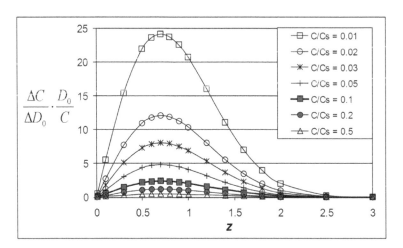

Figure 5.13 The sensitivity of the parameter D_0 in the prediction of the chloride concentration C.

The diffusion coefficient of D is relatively less sensitive if the C/C_s is larger than 0.1 and z value is larger than 1, the latter implying a thicker cover x or longer exposure time t.

Parameter k

The sensitivity coefficient of parameter k is similar to that of parameter D_0, i.e.

$$\frac{\partial C}{\partial k} = \frac{C_0}{\sqrt{\pi}} \frac{z}{k} e^{-z^2} \tag{5.20}$$

which results in an equation equivalent to Eq. 5.9:

$$\frac{\Delta C}{\Delta k} \frac{k}{C} = \frac{\Delta C}{\Delta D_0} \frac{D_0}{C} = \frac{1}{\sqrt{\pi}} \frac{z e^{-z^2}}{\dfrac{C}{C_s}} \tag{5.21}$$

Therefore, the relationships in Figure 5.13 are also valid for parameter k.

Parameter n' in Eq. 5.14

The sensitivity coefficient of parameter n' in Eq. 5.14 is

$$\frac{\partial C}{\partial n'} = \frac{C_0}{\sqrt{\pi}} z e^{-z^2} \ln\left(\frac{t'_0}{t}\right) \tag{5.22}$$

which results in

$$\frac{\Delta C}{\Delta n'} \frac{n'}{C} = \frac{n'}{\sqrt{\pi}} \frac{z e^{-z^2}}{\dfrac{C}{C_s}} \ln\left(\frac{t'_0}{t}\right) \tag{5.23}$$

Obviously, besides the influence of the C/C_s and z values, which is similar to that discussed above for the parameter D_0, $(\Delta C/\Delta n') \times (n'/C)$ is directly proportional to the value of n' and logarithmically proportional to the ratio t_0/t. In the case of $C/C_s = 0.1$ and $z = 1$, the relationships of $(\Delta C/\Delta n') \times (n'/C)$ and n' are shown in Figure 5.14.

Parameter n in Eq. 5.15

The sensitivity coefficient of parameter n in Eq. 5.15 is

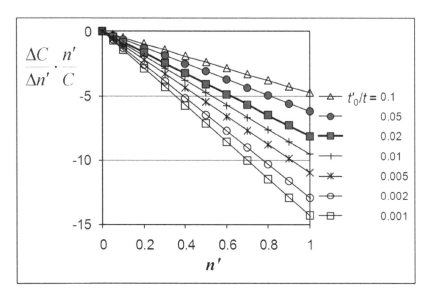

Figure 5.14 The sensitivity of the parameter n' in Eq. 5.14 for the prediction of the chloride concentration C: influence of the value n' when $C/C_s = 0.1$ and $z = 1$.

$$\frac{\partial C}{\partial n} = \frac{C_0}{\sqrt{\pi}} z e^{-z^2}$$

$$\left[\ln\left(\frac{t'_0}{t}\right) + \frac{1}{1-n} - \frac{\ln\left(1+\frac{t'_{ex}}{t}\right)\left(1+\frac{t'_{ex}}{t}\right)^{1-n} - \ln\left(\frac{t'_{ex}}{t}\right)\left(\frac{t'_{ex}}{t}\right)^{1-n}}{\left(1+\frac{t'_{ex}}{t}\right)^{1-n} - \left(\frac{t'_{ex}}{t}\right)^{1-n}} \right] \quad (5.24)$$

which results in:

$$\frac{\Delta C}{\Delta n}\frac{n}{C} = \frac{n}{\sqrt{\pi}}\frac{z e^{-z^2}}{\dfrac{C}{C_s}}$$

$$\left[\ln\left(\frac{t'_0}{t}\right) + \frac{1}{1-n} - \frac{\ln\left(1+\frac{t'_{ex}}{t}\right)\left(1+\frac{t'_{ex}}{t}\right)^{1-n} - \ln\left(\frac{t'_{ex}}{t}\right)\left(\frac{t'_{ex}}{t}\right)^{1-n}}{\left(1+\frac{t'_{ex}}{t}\right)^{1-n} - \left(\frac{t'_{ex}}{t}\right)^{1-n}} \right] \quad (5.25)$$

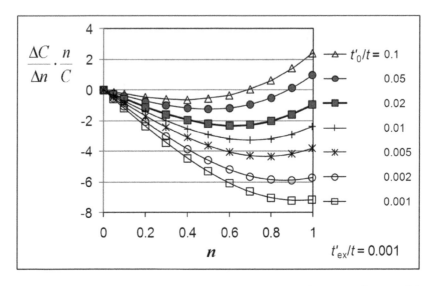

Figure 5.15 The sensitivity of the parameter n in Eq. 5.15 for the prediction of the chloride concentration C: influence of the value n when $C/C_s = 0.1$ and $z = 1$.

In this case, the effects of n and t'_0/t on $(\Delta C/\Delta n) \times (n/C)$ become complicated, and are illustrated in Figure 5.15, where $C/C_s = 0.1$ and $z = 1$.

Parameter t_0

The sensitivity coefficient of parameter t_0 in both Eq. 5.14 and Eq. 5.15 will be similar, i.e.

$$\frac{\partial C}{\partial t'_0} = \frac{C_0 n}{\sqrt{\pi}} \frac{ze^{-z^2}}{t'_0} \tag{5.26}$$

resulting in:

$$\frac{\Delta C}{\Delta t'_0} \frac{t'_0}{C} = \frac{n}{\sqrt{\pi}} \frac{ze^{-z^2}}{\dfrac{C}{C_0}} \tag{5.27}$$

Clearly, besides the influence of the C/C_s and z values, which is similar to that discussed above for the parameter D_0, $(\Delta C/\Delta t'_0) \times (t'/C)$ is directly proportional to the value of n. In the case of $C/C_s = 0.1$ and $z = 1$, $(\Delta C/\Delta t'_0) \times (t'/C) = 10/\sqrt{\pi} = 3.18n$.

5.3.3 The effect of different parameters on the sensitivity

In order to compare the effect of different parameters on the sensitivity, it is essential to make the parameter z constant, because this parameter is directly related to the chloride concentration C. For instance, in the case where $x = 0.05$ m and $t = 100$ years, it is not difficult to keep $z = 1$ through the adjustment of parameters D_0, n' or n.

Under the conditions $z = 1$ and $t'_0/t = 0.005$ (for parameter n only, corresponding to $t'_0 = 0.5$ years and $t = 100$ years), the effects of different parameters on the sensitivity can be summarised (Figure 5.16). It can be seen from Figure 5.16 that, of the different parameters, n or n' is the most sensitive, especially when its value is greater than 0.2. With the exception of parameter C_s, which is independent of the other parameters, the sensitivity of all the parameters is dependent on the ratio C/C_s.

5.3.4 Discussion

Under the typical exposure conditions, C_s may be about 5% of the binder content. If the threshold chloride content is 0.5% of binder content (e.g. in splash zone), C/C_s will be about 0.1. In this case, the sensitivity of parameters D_0 and any types of k will be 2 times as large as that of C_s, whilst the sensitivity of n, once it is larger than 0.2, will be more than 2 times as large as that of C_s. If the threshold chloride content is 1.5% of binder content

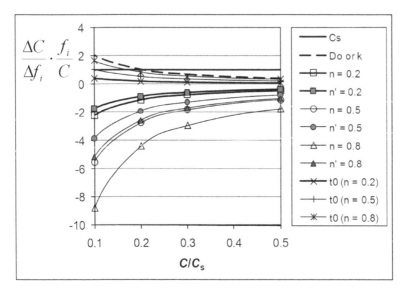

Figure 5.16 The effects of different parameters on the sensitivity under the conditions $z = 1$ and $t'_0/t = 0.005$ (for parameter n in Eq. 5.15 or n' in Eq. 5.14 only).

(e.g. in submerged zone), C/C_s will be about 0.3. Thus the sensitivity of parameters D_0 and any type of k will be less than that of C_s, but the sensitivity of n may still be larger than that of C_s.

5.3.5 Combined uncertainty of models for predicting chloride concentration

According to the *ISO Guide to the Expression of Uncertainty in Measurement*, ISO Guide 98–3 (2008), the combined standard uncertainty u_c is the positive square root of the combined variance u_c^2, which is given by

$$u_c^2 = \sum_{i=1}^{N} \left(\frac{\partial C}{\partial f_i} \right)^2 u_{f_i}^2 \tag{5.28}$$

where f denotes any type of parameter that may contribute to the uncertainty.

The coefficient of variation (COV) is defined as:

$$COV = \frac{u_c}{C} \tag{5.29}$$

The erfc model with constant D_a *and* C_{sa}

In the erfc model with constant D_a and C_{sa}, two parameters, C_s and D_0, will contribute to the uncertainty:

$$u^2 = \left(\frac{\partial C}{\partial C_s} \right)^2 u_{C_s}^2 + \left(\frac{\partial C}{\partial D_0} \right)^2 u_{D_0}^2 \tag{5.30}$$

$$COV = \sqrt{ \left(\frac{u_{C_s}}{C_s} \right)^2 + \frac{1}{\pi} \left(\frac{z e^{-z^2}}{\frac{C}{C_s}} \right)^2 \left(\frac{u_{D_0}}{D_0} \right)^2 } \tag{5.31}$$

Models with time-dependent $D_a(t)$ *and constant* C_{sa}, *or with time-dependent* $D_a(t)$ *and* $C_{sa}(t)$

In models with time-dependent $D_a(t)$ and constant C_{sa}, or with time-dependent $D_a(t)$ and $C_{sa}(t)$, four parameters, C_s, D_0, n and t_0, will contribute to the uncertainty:

$$u^2 = \left(\frac{\partial C}{\partial C_s} \right)^2 u_{C_s}^2 + \left(\frac{\partial C}{\partial D_0} \right)^2 u_{D_0}^2 + \left(\frac{\partial C}{\partial n} \right)^2 u_n^2 + \left(\frac{\partial C}{\partial t_0} \right)^2 u_{t_0}^2 \tag{5.32}$$

and

$$
\text{COV} = \sqrt{\left[\left(\frac{u_{C_s}}{C_s}\right)^2 + \frac{1}{\pi}\left(\frac{ze^{-z^2}}{\frac{C}{C_s}}\right)^2\right]\left\{\left(\frac{u_{D_0}}{D_0}\right)^2 + \left[\ln\left(\frac{t_0}{t}\right)u_n\right]^2 + n^2\left(\frac{u_{t_0}}{t_0}\right)^2\right\}} \quad (5.33)
$$

Model with time-dependent $D_a(t)$ and constant C_{sa}

In the model with time-dependent $D(t)$ and constant C_{sa}, seven parameters, $C_s, D_0, k_t, k_e, k_c, n$ and t'_0, will contribute to the uncertainty. The expressions are similar to those for models with time-dependent $D_a(t)$ and constant C_{sa} and models with time-dependent $D_a(t)$ and constant $C_{sa}(t)$:

$$
u^2 = \left(\frac{\partial C}{\partial C_s}\right)^2 u_{C_s}^2 + \left(\frac{\partial C}{\partial D_0}\right)^2 u_{D_0}^2 + \sum\left(\frac{\partial C}{\partial k_i}\right)^2 u_{k_i}^2 +
$$

$$
+ \left(\frac{\partial C}{\partial n}\right)^2 u_n^2 + \left(\frac{\partial C}{\partial t_0}\right)^2 u_{t_0}^2 \quad (5.34)
$$

and

$$
\text{COV}_{A1c} = \sqrt{\left[\left(\frac{u_{C_s}}{C_s}\right)^2 + \frac{1}{\pi}\left(\frac{ze^{-z^2}}{\frac{C}{C_s}}\right)^2\right]\left\{\left(\frac{u_{D_0}}{D_0}\right)^2 + \sum\left(\frac{u_{k_i}}{k_i}\right)^2 + \left[\ln\left(\frac{t_0}{t}\right)u_n\right]^2 + n^2\left(\frac{u_{t_0}}{t_0}\right)^2\right\}} \quad (5.35)
$$

where k_i represents k_t, k_e or k_c.

Apparently, the more parameters there are in a model, the more will be the sources of error. However, the actual COV for each model is dependent on the actual magnitude of each individual uncertainty. Besides, the most important factor for a prediction model is the degree of agreement with the actual chloride profiles.

5.3.6 Concluding remarks

- Parameter C_s has a constant sensitivity when predicting concentration, i.e. $(\Delta C/\Delta C_s) \times (C_s/C) = 1$, which is in some cases even larger than the sensitivity of parameters related to the diffusion coefficient (see below).

- The sensitivity of parameter D_0 or k (as a multiplier of D_0) is related to the ratio C/C_s and the value of z that contains all the variations except for C_s. If $z = 1$, when C/C_s is greater than 0.2, the sensitivity of D_0 or k when predicting concentration becomes less than parameter C_s, otherwise it becomes larger than C_s. However, when the value of z is very small or very large (not close to 0.7), the prediction of the chloride concentration becomes less sensitive to parameter D_0 or k.
- Of all the parameters, the age factor n or n' is the most sensitive, especially when its value is greater than 0.2.
- Parameter t_0 has a relatively lower sensitivity in the prediction of the chloride concentration, especially if the age factor n or n' is less than 0.5.

5.4 First comparison of predictions from early exposure data

The 'state-of-the-art' in modelling chloride ingress into concrete was tested at a Nordic (and *fib* TG) seminar in May 2001 (Nilsson, 2001). A test case was defined, with a given concrete of having given properties in a given environment. Exposure data for up to 2 years in the defined environment were given. Using those data it was easy to determine a suitable surface chloride content in that particular environment for the submerged zone.

A large number of model developers and model users were invited. Those who participated in the test and attended the seminar were from:

- Chalmers University of Technology, Göteborg, Sweden
- NTNU, Trondheim, Norway
- Lund University, Lund, Sweden
- University of Toronto, Canada
- Denmark's University of Technology, Lyngby, Denmark
- AEC A/S, Vedbaek, Denmark
- SINTEF, Trondheim, Norway
- FORCE, Bröndby, Denmark
- RWS, Utrecht, the Netherlands
- SP, Borås, Sweden
- Teknologisk, Taastrup, Denmark
- Imperial College, London, UK
- Elkem A/S, Kristiansand, Norway
- INSA, Toulouse, France
- Technische Universität, München, Germany
- Selmer-Skanska, Oslo, Norway
- Public Works Research Institute, Japan.

The participants were asked to use their preferred prediction model(s) to predict the chloride profiles after exposures of 30, 50 and 100 years, and the time to initiation of corrosion of reinforcing steel, in the test concrete in a selection of test environments.

5.4.1 The test concrete

An extensively investigated concrete in a well-known environment (from the Nordic project BMB, 'The Durability of Marine Concrete Structures') was chosen for the task of quantifying the differences between models. The mix composition is given in Table 5.8.

Various chloride migration and apparent diffusion coefficients have been determined for this concrete at different ages, as given in Table 5.9, where D_{nssm} is the non-steady-state migration coefficient determined using the rapid chloride migration (RCM) test (NT BUILD 492) and D_{nssd} is the non-steady-state diffusion coefficient determined using the immersion test (NT BUILD 443).

Specimens of this concrete have been exposed to seawater at the Träslövsläge field exposure site since 1992. The salinity and water temperature were $[Cl^-] = 14 \pm 4$ g/l and $T = 11 \pm 9°C$. The specimens were exposed at an age of 14 days.

Over a 2-year period cores were taken to determine the chloride profiles (Tang, 1997). The chloride profiles for the submerged zone of concrete H4 are shown in Figure 5.17. The 'achieved surface chloride content' C_{sa} was determined from these profiles by curve-fitting. The results are shown in Table 5.10.

Table 5.8 The composition of the selected test concrete (H4)

Water/ binder	Cement	Silica fume	Water	Cement paste	Binder content	Calc. density	Air
0.40	399.0 kg/m³	21.0 kg/m³	168.0 kg/m³	30.4 vol %	19% mass	2210 kg/m³	5.9 vol %

Notes
Cement: SRPC. Silica fume: slurry.

Table 5.9 Diffusion coefficients for the test concrete (H4) obtained from accelerated tests (Tang, 1997)

D_{nssm} (10⁻¹² m²/s)				D_{nssd} (10⁻¹² m²/s)
Unexposed (age 0.5 years)	Exposed (age 1 year)	Unexposed (age 1.3 years)	Exposed (age 2 years)	Unexposed (age 0.5 years)
2.7	3.2	3.7	3.1	1.7

Table 5.10 The achieved surface chloride content C_{sa} for concrete H4 after a 2-year exposure at Träslövsläge (Tang, 1997)

Age (years)	0.63	1.05	2.03						0.6–2
Exposure time (years)	0.59	1.01	1.99						Average
Weight % of binder	3.82	3.26	3.71	3.66	3.75	3.95	3.44	3.45	3.63

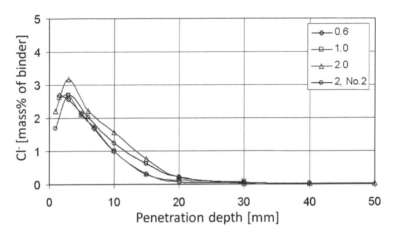

Figure 5.17 Chloride profiles in the submerged zone for concrete H4 during a 2-year field exposure period (Tang, 2003a).

A concrete cover of 60 mm was assumed in all the environments considered (see below).

5.4.2 *The test environments*

Case A: *bridge column in a marine environment*

The annual average ± variation of the salinity and water temperature were: $[Cl^-] = 14 \pm 4$ g/l, $T = 11 \pm 9°C$. The concrete was assumed to be exposed at an age of 14 days.

- Marine environment 1: submerged zone.
- Marine environment 2: tidal zone.
- Marine environment 3: splash zone.
- Marine environment 4: atmospheric zone.

Case B: *bridge column in a road environment*

Road environment 1: A column at a distance from the road lane of 3 m was considered. An average total of 0.15 kg/m² NaCl was assumed to be spread during the five 'winter' months each year. Other required climatic conditions were appropriately assumed.

5.4.3 *The test results*

The predictions of the chloride ingress into the concrete in the marine submerged and splash zones, as obtained using the various models, are compared in Figures 5.18 to 5.21. Significant scatter was found in predictions of further ingress, even by the 'best' prediction models available, and even for the submerged zone. For the splash zone the scatter was substantial. These results clearly show that there is a long way to go before chloride-ingress modelling is accurate enough for design applications.

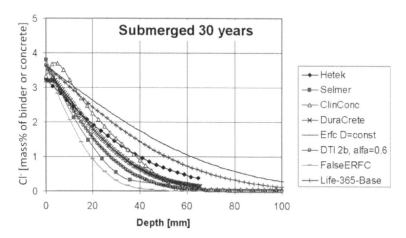

Figure 5.18 Chloride profiles obtained using various prediction models for a given concrete in a well-defined environment. Profiles for exposures of up to 2 years given. Prediction results for 30 years in the submerged zone (Nilsson, 2001).

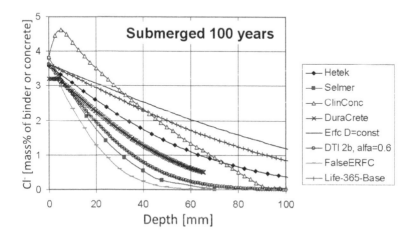

Figure 5.19 Chloride profiles obtained using various prediction models for a given concrete in a well-defined environment. Profiles for exposures of up to 2 years given. Prediction results for 100 years in the submerged zone (Nilsson, 2001).

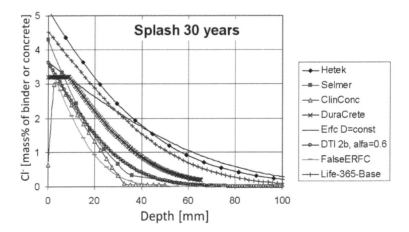

Figure 5.20 Chloride profiles obtained using various prediction models for a given concrete in a well-defined environment. Profiles for exposures of up to 2 years given. Prediction results for 30 years in the splash zone (Nilsson, 2001).

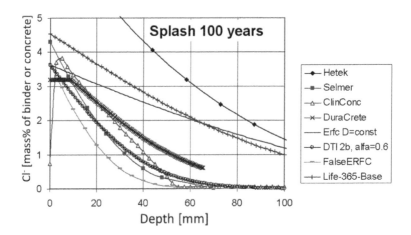

Figure 5.21 Chloride profiles obtained using various prediction models for a given concrete in a well-defined environment. Profiles for exposures of up to 2 years given. Prediction results for 100 years in the splash zone (Nilsson, 2001).

The differences between the predictions obtained may be explained by the background exposure data for each prediction model, which data were more or less individual for each model. Better data and better exchange of data would probably reduce the scatter significantly. This exercise remains to be done.

5.5 Second comparison of predictions from early exposure data

5.5.1 *Objectives and overview of work performed*

In the EU CHLORTEST project a new benchmarking of models was done by the partners, IETcc in Spain (also task leader), SP and LTH in Sweden, and EDF in France, as a subtask in the work package for modelling chloride ingress. The objective was to study the ability of models to reproduce reality, with laboratory test results as input and in-field performance data as the comparison. This was done by selecting some field data, for which the composition and properties of the concrete and the environmental conditions were well documented, and asking model developers to make predictions using their models on these field cases. The prediction results were then analysed using a probabilistic approach.

The complete background data and prediction results were presented in a final working report by the task leader, IETcc (Nilsson, 2005). This report was complemented with an amendment by the work package leader LTH, which added and corrected some data and analysis. The various stages involved in the benchmarking analysis were:

1   the establishment of the criteria for the benchmarking;
2   the selection of profiles from the database;
3   the selection of model producers with a worldwide reputation;
4   the submission of the documentation to the modellers;
5   the collection of results and the application of the benchmarking criteria;
6   discussion of the results obtained and suggestions for further action.

Each of these stages is elaborated further in the following subsections.

5.5.2 *Establishment of criteria for benchmarking*

Before commencing the work, the criteria for making the comparison between predicted and real profiles were established by IETcc and EDF in a collaborative effort. The method proposed for undertaking the benchmarking was selected on the basis of a comparison of the bias of the predicted profiles obtained with each model with respect to the real profile, in terms of the areas between the two profiles, as explained below. This approach has the advantage of quantifying, to some degree, the deviation of the predictions, and therefore allowing comparisons to be made between models.

Comparison of different models when the measured data are available

In order to verify the performance of the different mathematical approaches used by different researchers, a methodology based on the area between

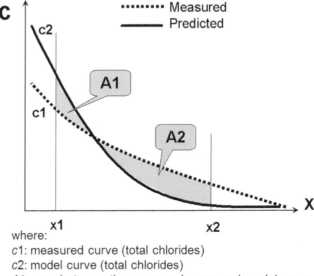

where:
c1: measured curve (total chlorides)
c2: model curve (total chlorides)
A1: area between the measured curve and model curve,
    for which measured data is lower (negative sign)
    [%·mm]
A2: area between the measured curve and model curve,
    for which measured data is higher (positive sign)
    [%·mm]
x1, x2: validation depth range
C, X: concentration and depth axis respectively

Figure 5.22 Predicted and measured profiles of chloride concentration.

the measured and the predicted curves was developed (Figure 5.22). The prediction result obtained with each model can be compared with the measured curve by comparing the value of the area $(A_1$ and $A_2)$ between the two curves over a depth range from x_1 to x_2, as shown in Figure 5.22.

The values used for the comparison were the areas A_1 and A_2 with and without a sign, respectively:

$$S_1 = |A1| + |A2| \qquad\qquad (5.\ 36{:}1)$$

and

$$S_2 = |A1| + |A2| \qquad\qquad (5.\ 36{:}2)$$

Further information includes:

- The curves c_1 and c_2 are cubic spline interpolations of the measured and model data points, respectively. If the measured data or predicted profile did not cover the whole interval between x_1 and x_2 an extrapolation was made.
- S_1 indicates how close the modelled profile is to the measured data.
- S_2 indicates how much higher or lower the predicted curve is compared with the measured data.

$S_1 > 0$
$S_2 < S_1$ at all times
$S_2 < 0$ if the measured profile is lower than the predicted one (overestimation)
$S_2 > 0$ if the measured profile is higher than the predicted one (underestimation)
$S_2 = -S_1$ if the whole of the measured profile is lower than the predicted one
$S_2 = S_1$ if the whole of the measured profile is higher than the predicted one.
The validation range adopted was:

- $x_1 = 10$ mm (this value was adopted to eliminate the skin effect);
- $x_2 = 50$ mm (except in the situation when the depth of the measured data is smaller).

It is obvious from some of the reported predictions that this choice of validation range is questionable where the depth of penetration is small. In that case, a significant part of the profile falls between 0 and x_1, which is excluded from the comparison.

Comparison of different models when measured data are unavailable

When measured data are unavailable, such as predicted chloride profiles at 100 years, the table containing the modellers' data and graphs are given, but no comparison is made.

5.5.3 Selection of profiles and documentation prepared for benchmarking

The purpose is to know the deviation of the predicted data from the real data. This deviation was evaluated by using a supplied real profile at one concrete age to predict a second profile of the same concrete at a second age. In addition, a prediction at 100 years was selected in order to evaluate the sensitivity of the models to take into account the effect of concrete age on diffusivity.

Therefore, the in-field chloride profiles used for the verification of the modelling were required to meet the following criteria:

- The concrete mix proportions are known.
- The laboratory test methods and results for diffusivity are known.
- The exposure conditions are known.
- The samples for determining chloride profiles are obtained under the same conditions at, preferably, three concrete ages: one or two at concrete ages of less than 2 years (for estimating the age effect), and one or two at concrete ages of more than 10 years (for verification of the long-term prediction).

As the task consisted of studying the ability of the different models to predict the future evolution of the chloride profiles, the prediction being made using the results (a profile) obtained at one concrete age, accompanied or not by short-term results obtained in the laboratory, of the same concrete. The profiles to be used in the task were selected as follows:

1 A set of chloride profiles from the database of the CHLORTEST project was selected following the criteria given above:

   - A set of 400 profiles was collected and passed on to the task group.
   - It was recognised that most of the profiles did not follow the criteria fixed previously by SP and LTH.
   - Only profiles provided by SP met all the criteria.
   - It seemed to the task group that a wider variety of profiles is necessary to ensure sufficient objectivity of the benchmarking exercise.

2 A brief guide was prepared and sent to the modellers. The guide contained:

   - a summary of the project;
   - the purpose of the benchmarking exercise as the subtask;
   - the three sets of profiles, with instructions regarding the predictions required and forms for submitting results;
   - a form to be returned to indicate whether the modeller was willing to collaborate in the task;
   - a questionnaire for modellers to specify the characteristics of the models used;
   - Excel files containing the given profiles and space to insert the predicted ones.

The three sets of profiles that were selected corresponded to 22 different cases. The profiles were supplied to the modellers, with instructions for the required predictions, as follows:

- Series 1: one earlier age profile of two chloride profiles from two consecutive ages for the same concrete was supplied to the modeller. Results of short-term accelerated tests were also included. The modeller was asked to:
    - predict the second age profile
    - predict the profile at 100 years.

- Series 2: one earlier age profile of two chloride profiles from two consecutive ages for the same concrete was supplied to the modeller. No results of short-term accelerated tests was available. The modeller was asked to:
    - predict the second age profile
    - predict the profile at 100 years.

- Series 3: one profile from an old structure was given. The modeller was asked to:
    - predict the profile at 100 years.

The modellers were encouraged to model all the cases if possible, but they could model a limited number of cases if their time was limited:

- For Series 1, the exercise was compulsory for the first four profiles supplied. The remainder of the profiles in Series 1 could be treated as voluntary.
- For Series 2 and 3, the exercise was compulsory for the first profile. The remainder of the profiles in these series could be treated as voluntary.

In Series 1 profiles the concrete chloride-ingress properties were given as a measured chloride migration coefficient. This kind of information was not available in Series 2 or 3. The Series 1 profiles supplied to the modellers were for a very short time of exposure, this being less than 10% of the exposure time for the required prediction. The Series 2 and 3 profiles, however, were for exposure times representing about 40% of the required prediction.

The 22 cases represented in the profiles are summarised in Table 5.11.

5.5.4 Selection of models

The selection was made by considering the following two criteria:

- models developed by individuals which were published in the scientific literature and had become popular;
- models which were well known and were being used, e.g. 'error function' or 'square root' models.

Table 5.11 Summary of the characteristics of the 22 cases represented in the three profile series

Case	Concrete	Environment	Exposure time (years)		Comments
			Given profile	Measured/ predicted	
Series 1					
C1	CEM I	Submerged	0.8	10.3	Limited thickness
C2	w/c = 0.4	Splash			
C3		Atmospheric			
C4	CEM I w/c = 0.35	Road: vertical surface	0.4	4.6	Small difference: 0.4–4.6 years
C5	CEM I	Submerged	1.0	10.2	
C6	w/b = 0.35 5% SF 20% FA			10.1	
C7	CEM I w/c = 0.4	Road: horizontal surface	0.4	4.6	Small difference: 0.4–4.6 years
C8	CEM I	Submerged	0.7	10.2	
C9	w/b = 0.35 5% SF 10% FA	Splash			
C10	CEM I	Atmospheric	0.5	3.0	
C11	w/c = 0.5	Tidal			
C12	CEM I w/c = 0.4 5% SF	Road: vertical surface	0.4	4.6	Small difference: 0.4–4.6 years
C13		Road: horizontal surface			
Series 2					
C1	OPC	Tidal	7	12	Small difference: 6–14 years
C2	SRPC CEM I w/c = 0.5		6	14	
C3	CEM II/B w/c = 0.57	Splash	8	18	Small difference: 8–18 years
C4	CEM IV/B	Submerged	None	1.5	
C5	w/c = 0.4	Tidal			
Series 3					
C1	CEM III/B w/c = 0.45	Tidal: 1 m above sea	42	None/100	
C2		Atmospheric: 9 m above sea			
C3		Atmospheric: 14 m above sea			
C4					No real difference

Note
FA, fly ash; SF, silica fume; w/b, water/binder ratio; w/c, water cement ratio, OPC, ordinary Portland cement, SRPC, sulphate-resistant Portland cement.

A total of 18 model developers from different countries were invited to partake in the benchmarking evaluation.

5.5.5 Responses obtained

Not all the model developers could, for different reasons, participate in the benchmarking evaluation. Finally, 16 models (Table 5.12) from nine different model developers were considered in the evaluation. These 16 models were thought to be sufficiently representative for the purpose of benchmarking. All the models were kept anonymous by using model codes (Model X) in the evaluation, except for model Model 4, which is the traditional erfc model.

5.5.6 Comparison of results

The results for Series 1 were presented in the form listed below for each case:

- a table containing the predicted and the real profile at the second age;
- a graph showing the predicted and real profile at the second age;
- a histogram of the S_1 value found by the modellers for the second age;
- a histogram of the S_2 value found by the modellers for the second age;
- comments about the results obtained for the second age;
- a table containing the predicted and the real profile for the 100 year predictions;
- a graph showing the predicted and the real profile for the 100 year predictions;
- comments about the results of the 100 year predictions.

All results were documented in the final working report of the CHLORT-EST project (Andrade, 2005). Some examples are given in Section 5.5.7, and an analysis of all the predictions is given in Section 5.5.8.

5.5.7 Some examples of results

Series 1, Case 1: submerged seawater environment

Series 1, Case 1 is a CEM I concrete with a water/cement ratio of 0.40, constantly submerged in seawater. The given profile after an exposure time of 0.78 years is shown in Figure 5.23. This profile has a surface chloride content C_s of 2.5–3% of the binder content.

The predicted profiles after an exposure time of 10.3 years are shown in Figure 5.24 together with the measured profile (for Model 0). The differences between the predicted and the measured profiles, expressed as S_1 and S_2 values, are shown in Figures 5.25 and 5.26, respectively.

Table 5.12 Models participated in the benchmarking comparison test

Model	Case																					
	Series 1													Series 2					Series 3			
	C1	C2	C3	C4	C5	C6	C7	C8	C9	C10	C11	C12	C13	C1	C2	C3	C4	C5	C1	C2	C3	C4
1	1, 2	1, 2	1, 2	1, 2	1, 2	1, 2	1, 2	1, 2	1, 2	1, 2	1, 2	1, 2	1, 2	1, 2	1, 2	1, 2	1, 2	1, 2	1	1	1	1
2	1, 2	1, 2	1, 2	1, 2	1, 2	1, 2	1, 2	1, 2	1, 2	1, 2	1, 2	1, 2	1, 2									
3	1, 2	1, 2	1, 2	1, 2	1, 2	1, 2	1, 2	1, 2	1, 2	1, 2	1, 2	1, 2	1, 2									
4	1, 2	1, 2	1, 2	1, 2	1, 2	1, 2	1, 2	1, 2	1, 2	1, 2	1, 2	1, 2	1, 2	1, 2	1, 2	1, 2	1, 2		1	1	1	1
5	1, 2	1, 2	1, 2	1, 2	1, 2	1, 2	1, 2	1, 2	1, 2	1, 2	1, 2	1, 2	1, 2	1, 2	1, 2	1, 2	1, 2	1, 2	1	1	1	1
6	1, 2	1, 2	1, 2	1, 2	1, 2	1, 2	1, 2	1, 2	1, 2	1, 2	1, 2	1, 2	1, 2	1, 2	1, 2	1, 2			1	1	1	1
7	1, 2	1, 2	1, 2	1, 2	1, 2	1, 2								1, 2								
8	1, 2	1, 2	1, 2	1, 2	1, 2	1, 2								1, 2								
9	1, 2	1, 2	1, 2	1, 2	1, 2	1, 2	1, 2	1, 2	1, 2	1, 2	1, 2	1, 2	1, 2	1, 2	1, 2	1, 2			1	1		
10	1, 2	1, 2	1, 2	1, 2	1, 2	1, 2	1, 2	1, 2	1, 2	1, 2	1, 2	1, 2	1, 2	1, 2	1, 2	1, 2			1	1		
11				1, 2			1, 2				1, 2	1, 2	1, 2									
12	1, 2	1, 2	1, 2	1, 2	1, 2	1, 2	1, 2	1, 2	1	1, 2	1, 2	1, 2	1, 2						1	1	1	1
13	1, 2	1, 2	1, 2	1, 2	1, 2	1, 2	1, 2	1, 2	1, 2	1, 2	1, 2	1, 2	1, 2						1	1	1	1
14																						
15	1, 2	1, 2	1, 2		1, 2	1, 2		1, 2	1, 2	1, 2	1, 2			1, 2	1, 2	1, 2			1	1	1	1
16	1, 2	1, 2	1, 2		1, 2	1, 2		1, 2	1, 2	1, 2	1, 2			2	1		2					

Notes
1 The profile at the second age was produced by the model developer.
2 The profile at 100 years was produced by the model developer.

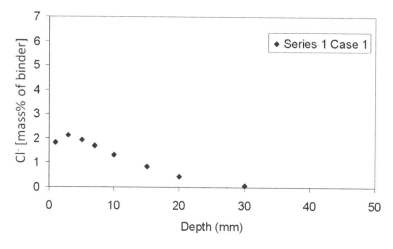

Figure 5.23 Series 1, Case 1: the given profile after an exposure of 0.78 years.

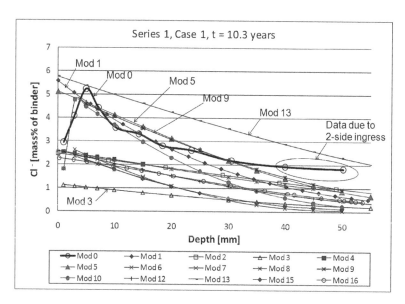

Figure 5.24 Series 1, Case 1: modelled profiles for 10.3 years of exposure. Mod 0 is the real measured profile.

COMMENTS ON SERIES 1, CASE 1, 10.3 YEARS

First, it should be noted that the profile at 10.3 years does not represent a semi-infinite case. The profiles were taken from concrete slab specimens with a thickness of only 100 mm and penetration of chlorides from two

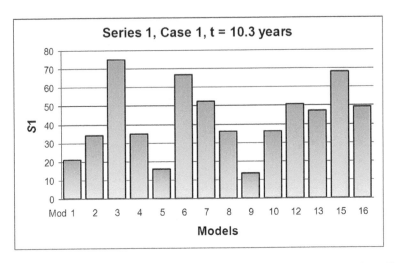

Figure 5.25 Series 1, Case 1: the area S_1 between the real and the predicted profiles.

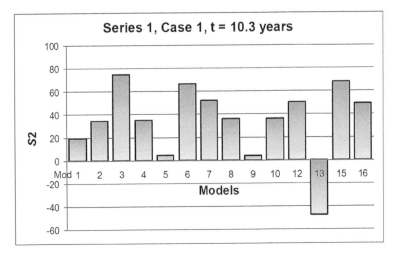

Figure 5.26 Series 1, Case 1: the area S_2 between the real and the predicted profiles.

sides. For an exposure time of 10.3 years, the chloride penetration depth from each surface is greater than 50 mm. Consequently, the measured chloride profile is not fully relevant for comparing with the predicted profiles, because all predictions are made for semi-infinite cases. Bearing this in mind, the results could be analysed with some limitations.

There are large differences between the predicted profiles. Roughly, some models (Model 1, 5, 9, 10 and 13) correctly predicted the change in C_s with time, and therefore they fit the measured profile much better. Others, not having selected the new real C_s, used instead a constant C_s value, and the predicted profiles in these cases are very far from the measured one in the initial part of the profile near the surface.

There are also large differences in the predicted profiles between models using a time-dependent or constant C_s value. Even though the input data included a given, 'early' profile and a value of D, the actual values of D used in the predictions are very different.

However, it is interesting to realise that there are no large differences in the 'front part' of the predicted profiles between those models that selected, more or less well, the new C_s and those that considered a constant C_s. That is, at 5 cm depth the predicted profiles do not differ too much. This is, of course, due to the low D in all cases.

In summary, despite the difficulties associated with the prediction, some very good predictions were obtained in this case. These are mainly those obtained using Models 5 and 9.

COMMENTS ON SERIES 1, CASE 1, 100 YEARS

Here the prediction is up to 100 years (Figure 5.27). The prediction at 5 cm depth of course differs much more between the models that consider a time-dependent increase in C_s and those that did not consider this. Therefore, the C_s change has a strong influence on the long-term prediction. There is also a significant difference in the D values used.

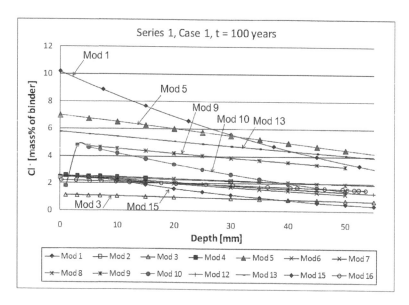

Figure 5.27 Series 1, Case 1: modelled profiles for 100 years of exposure.

Series 1, Case 4: road environment

Series 1, Case 4 involves a similar concrete as in Case 1, but exposed in a road environment. The given profile was measured after 0.4 years, i.e. during the first year (see Figure 5.28). Actually, the profile was taken just after the first winter, before the rains in the summer and autumn, when chloride is leached out.

The predicted profiles after 4.6 years are shown in Figure 5.29 together with the measured one (Model 0). All predicted profiles overestimated the

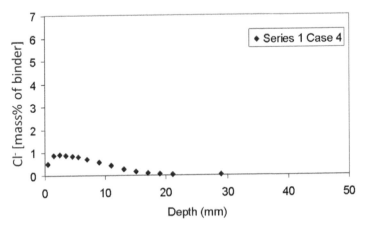

Figure 5.28 Series 1, Case 4: the given profile after an exposure of 0.4 years.

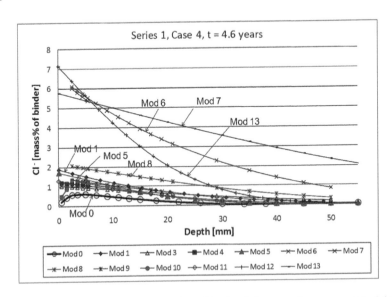

Figure 5.29 Series 1, Case 4: modelled profiles for 4.6 years of exposure. Model 0 is the real measured profile.

chloride contents after 4.6 years, whereas the profile measured at 4.6 years actually shows a lower chloride content in the outer part of the concrete than the measured profile after 0.4 years.

The measured profile shows very low C_s values, and therefore those models that consider a constant C_s value equal to the given one produced more accurate predictions than those that assumed an increase in C_s with time. The latter models failed strongly in their predictions.

The models that achieved a prediction closest to the measured profile are Models 3, 8 and 10.

5.5.8 Analysis of all predictions

Series 1

The other cases in Series 1 gave results similar to those for the cases presented above. The S_1 values for all the cases in Series 1 are shown in Table 5.13 and the S2 values in Table 5.14. The tables also give the average S1 and S2 value and the standard deviation for each model and each case.

In some cases (Models 5, 9 and 10) there was excellent agreement between the predicted and the measured profiles, although a few of the models predicted profiles which are far from the measured ones. Models 1 and 2 gave very poor results for most of the cases. The extreme increase in C_s with time from 0.8 years to 10 years was difficult to predict.

One model, Model 5, stood out in this comparison. This model is, in fact, the ClinConc model (see Section 4.5.3), which has previously been adjusted using most of the data (Swedish data) in Series 1, and thus it is not surprising that this model gave the best prediction of all the models. Four more models form a second group, Models 1, 9, 10 and 12. These models gave predictions that are far from the measured profiles for some of the cases.

However, this means of comparing models is doubtful. The magnitude of the S_1 and S_2 depends on the magnitude of the chloride content for each profile. A profile with high chloride concentrations will give a much higher S_1 value than will a profile with low chloride contents. Consequently, the different cases will contribute differently to the comparison. The comparison method is good for comparisons between different models analysing the same case, but its use is questionable for comparisons between different cases. Thus the S_1 and S_2 values should be normalised, for instance by giving them as a percentage of the area of the measured profile, in order to give better comparison between different cases. Alternatively, the models could be ranked according to their S_1 values for each case, which would give an indication of closeness of the modelled profile to the measured one. S_2 values cannot be used for ranking, but only for indicating overestimation ($S_2 < 0$) or underestimation ($S_2 > 0$). A summary of the ranking of the models for all the cases in Series 1 is given in Table 5.15. Clearly Model 5 gave the prediction that was closest to the measured data.

Table 5.13 Overview of all predictions for Series 1, compared with the measured profiles, expressed as S_1 values

Model	Case													Average	SD
	C1	C2	C3	C4	C5	C6	C7	C8	C9	C10	C11	C12	C13		
1	21.2	50.3	9.1	12.1	5.0	16.7	12.8	30.7	5.6	7.1	17.9	6.9	3.8	15.3	13.0
2	34.2	63.1	11.0		8.4	25.4		21.1	8.4					24.5	19.6
3	75.2	69.7	11.6	13.4	10.2	21.0	15.1	21.4	7.0	21.8	25.0	10.1	12.9	24.2	22.1
4	35.0	65.0	9.6	15.8	6.6	24.5	14.9	20.4	8.0	6.1	16.7	16.5	20.4	20.0	15.7
5	15.8	12.4	1.6	6.1	4.3	6.8	7.8	6.5	1.9	5.8	9.9	2.5	1.5	6.4	4.3
6	66.8	46.3	49.3	63.8	9.9	8.6	59.2	18.8	27.7	5.0	7.7	53.0	59.3	36.6	24.0
7	52.4	25.3	76.6	89.4	7.9	10.2								43.6	34.6
8	36.3	65.5	13.5	30.8	6.2	23.4								29.3	20.9
9	13.7	72.3	4.2	13.0	7.9	18.6	12.7	18.4	3.3	6.4	18.0	8.8	8.3	15.8	17.7
10	36.2	74.4	2.8	6.5	3.0	10.5	3.3	9.5	8.7	6.0	22.0	2.5	5.5	14.7	20.3
11				22.5			18.9					24.1	30.9	24.1	5.0
12	50.5	76.4	3.1	6.8	9.4	9.2	7.4	20.6	2.3	7.5	17.7	3.9	7.8	17.1	21.9
13	47.3	44.4	19.3	44.6	4.7	9.0	46.3	19.6	4.6	4.6	29.3	10.4	12.8	22.8	17.3
14															
15	68.2	84.4	2.7		12.4	10.7		19.1	2.3	10.7	19.3			25.5	29.7
16	49.4	69.1			6.3	21.9		16.6			12.5			29.3	24.6
Average	43.0	58.5	16.5	27.1	7.3	15.5	19.8	18.6	7.3	8.1	17.8	13.9	16.3		
SD	19.1	20.4	22.0	26.3	2.6	6.8	18.2	6.0	7.2	5.1	6.3	15.3	17.4		

Notes

The shading in the table give some additional information, compared to the final task report by Andrade (2005):

▮ Data missing; now calculated and added.

▮ Data wrong; now corrected.

▯ Best comparison.

SD, standard deviation

Table 5.14 Overview of all predictions for Series 1, compared to the measured profiles, expressed as S_2 values

Model	Case													Average	SD
	C1	C2	C3	C4	C5	C6	C7	C8	C9	C10	C11	C12	C13		
1	19.5	50.3	-9.1	-12.1	-5.0	16.7	-12.8	30.7	-5.6	-7.1	16.0	6.9	3.8	7.1	18.9
2	34.2	63.1	-11.0		3.8	-25.1		-4.2	-8.3	-21.8	-12.0	-9.7	-12.9	7.5	30.6
3	75.2	69.7	-11.6	-13.4	8.1	21.0	-15.1	7.9	-6.6	0.7	-0.7	-16.5	-20.4	6.1	31.7
4	35.0	65.0	-9.6	-15.8	1.2	-24.5	-14.9	-3.8	-8.0	-0.7	0.6	2.5	-0.6	-0.9	24.8
5	4.3	-1.1	1.3	-6.1	-0.6	-5.8	-7.8	-3.4	1.3	-5.0	-2.1			-1.2	3.6
6	66.8	46.3	-49.3	-63.8	7.9	4.1	-59.2	17.5	-27.7			-53.0	-59.3	-13.6	42.5
7	52.4	25.3	-76.6	-89.4	5.8	-6.8								-14.9	56.6
8	36.3	65.5	-13.5	-30.8	2.1	-23.2	-12.7	-15.7	-3.3	1.4	2.5	-8.8	-8.3	6.1	37.6
9	4.1	72.3	-3.3	-13.0	7.9	-18.6	-2.8	2.9	-8.7	2.9	20.5	-1.4	-5.1	0.4	23.1
10	36.1	74.4	-0.9	-6.2	1.6	10.4								9.5	23.0
11				-22.5			-18.9					-24.1	-30.9	-24.1	5.0
12	50.5	76.4	-2.7	-6.8	9.4	7.9	-7.4	20.4	2.2	7.4	13.1	3.6	-7.6	12.2	24.9
13	-47.3	44.4	-19.3	-44.6	4.5	8.9	-46.3	19.6	4.6	-3.6	29.3	-6.6	-9.7	-5.1	28.7
14															
15	68.2	84.4	2.7		12.4	10.7		19.1	2.3	10.7	17.7			25.4	29.7
16	49.4	69.1			1.9	-16.6		4.5			5.1			18.9	32.9
Average	34.6	57.5	-15.6	27.0	4.4	-2.9	-19.8	8.0	-5.3	-1.5	8.2	-11.4	-15.1		
SD	32.1	22.9	22.6	26.3	4.6	16.3	-18.2	13.6	8.9	8.9	12.1	17.1	18.4		

Notes

The shading in the table give some additional information, compared to the final task report by Andrade (2005):

Data missing; now calculated and added.

Data wrong, now corrected.

Best comparison

SD, standard deviation

Table 5.15 Model ranking based on the S_1 values obtained for all the cases in Series 1

Model	Case C1	C2	C3	C4	C5	C6	C7	C8	C9	C10	C11	C12	C13	Mean
1	3	5						11	5	7	6			6.2
2	4	6	6		12			9	8					7.5
3	14	10	7	5	11	8	6	10	6	10	10	5	6	8.3
4	5	7	5	6	6	11	5	7	7	5	4	7	7	6.3
5	2	1	1	1	2	1	3	1	1	3	2	2	1	1.6
6	11	4	10	10	10	2	9	5	10	2	1	9	9	7.1
7	13	2	11	11	7	5								8.2
8	7	8	8	8	4	10								7.5
9	1	11	4	4	8	7	4	4	3	6	7	4	4	5.2
10	6	12	2	2	1	6	1	2	9	4	9	1	2	4.4
11				7			7				8	8		7.5
12	10	13	3	3	9	4	2	8	2	8	5	3	3	5.6
13	8	3	9	9	3	3	8	6	4	1	11	6	5	5.8
14														
15	12									9	8			9.7
16	9	9			5	9		3			3			6.3

Series 2

A comparison of the measured and predicted profiles obtained for the cases in Series 2 are shown in Tables 5.16 and 5.17. Some of the predicted profiles were so unexpectedly low that they have not been quantified. These erroneous profiles were obtained because there was a mistake in the profiles given to the modellers (expressed as a percentage by weight of concrete instead of

Table 5.16 Overview of all predictions for Series 2, compared to the measured profiles, expressed as S_1 values

Model	Case C1	C2	C3	C4	C5	Average	SD
1	29.4	42.4	94.4	8.7	10.2	37.0	35.0
4	45.3	34.6	8.5			29.4	19.0
5	12.7	11.9	7.9	11.3	16.0	12.0	2.9
6	9.0	9.0	11.8			9.9	1.6
7	13.3					13.3	
8	10.3					10.3	
9	10.6	11.5	11.3			11.1	0.5
10	18.9	7.9	3.3			10.0	8.0
15	×	×	4.5			4.5	
16		26.6				26.6	
Average	18.7	20.6	20.2	10.0	13.1		
SD	12.6	13.9	32.9	1.8	4.1		

First x_1 = 15 mm   First x_1 = 20 mm

Notes

▉ Data missing, now calculated and added.

× Data missing, based on original erroneous data.

SD, standard deviation.

Table 5.17 Overview of all predictions for Series 2, compared to the measured profiles, expressed as S_2 values

Model	C1	C2	C3	C4	C5	Average	SD
1	−3.6	−41.5	−94.4	−8.7	0.1	−29.6	39.8
4	45.3	34.6	−8.4			23.8	28.4
5	0.0	−11.8	−7.9	−11.3	16.0	−3.0	11.6
6	1.5	−9.0	−11.8			−6.4	7.0
7	8.2					8.2	
8	10.3					10.3	
9	3.5	−11.5	−10.9			−6.3	8.5
10	13.3	−3.8	2.9			4.1	8.6
15	×	×	−2.7			−2.7	
16		26.6				26.6	
Average	9.8	−2.3	−19.0	−10.0	8.1		
SD	15.4	25.7	33.6	1.8	11.2		

First $x_1 = 1.5$mm
First $x_1 = 20$mm

Notes
█ Data missing, now calculated and added.
× Data missing, based on original erroneous data.
 Best Comparison
SD, standard deviation.

as a percentage of binder content). Some of the model developers did not change their predictions after new profiles were distributed.

The number of predictions was small for cases 4 and 5, and therefore these cases are not considered further.

Once again, Model 5 gave good predictions, as did Models 9 and 10. The cases in Series 2 are 'new' for Model 5, contrary to the cases in Series 1 (see above). Other models gave some good predictions, but dealt only with one or two cases.

Series 3

In Series 3 the given profiles were obtained at 42 years as a 'calibration' of the models. Despite this, the predicted profiles at 100 years of exposure obtained using the different models showed significant differences.

One example is shown in Figure 5.30 for Case 1 in Series 3. Case 1 is a concrete exposed in the tidal zone, which should be 'easier' than Cases 2–4, which are concretes in the atmospheric zone.

The given profile obtained at 42 years had a C_s value of 2.5% of the binder content. However, the predicted surface chloride contents vary from over around 4.0%, up to 12% and even above 30%. It is also clear from the profiles that the prediction models use very different diffusion coefficients, even though a calibration was possible after 42 years.

5.5.9 Final comments on the benchmarking evaluation of models

The most important observations regarding the benchmarking exercise are summarised below.

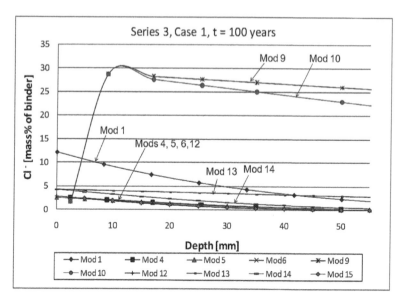

Figure 5.30 Series 3, Case 1: modelled profiles for 100 years of exposure. The given profile was for 42 years of exposure, as 'calibration'.

- With regard to the sensitivity of the models to the effect of their constituent parameters, it can be deduced that the parameter that has the strongest influence is the C_s value, in both the short- and long-term (100 years) predictions. This parameter was also identified in Section 5.3.6. The results of the benchmarking evaluation confirm the importance of including correct assumptions regarding the boundary conditions if reliable predictions are to be achieved.
- Almost all the models have C_s as the input, apart from a few models that have the free c as the input and convert this parameter to C_s by using binding factors. The finding from the benchmarking exercise of a relatively large influence of C_s and a relatively less influence of D indicates that the assumption of a constant C_s, as used in most models, is far from the real situation.
- Better models for predicting the effects of the environment on concrete are urgently required.
- With regard to the D values, these have a lesser influence than the C_s values on the whole profile, but more significantly influence the chloride penetration front, as also identified in Section 5.3.6. It should be mentioned that D values were given to the model developers only in some of the profiles supplied in Series 1, which means that model developers used the D values they considered to be the most convenient in the other two profile series.

- Where the same surface chloride conditions were used, very different D values were used in different models, even though an initial profile and a measured D value was given as input data. The model developers obviously treated chloride diffusion coefficients in very different ways.

- Finally, with regard to the methodology used to undertake the benchmarking evaluation, although the cases could have been better chosen and the analysis could have been more thorough, the evaluation has been very useful with regard to comparing measured and predicted profiles.

5.6 Validation against long-term exposure data

5.6.1 Data collected over 10 years of exposure in a marine environment

In the early 1990s, as part of a Swedish national project, BMB, 'Durability of Marine Concrete Structures' (Sandberg, 1996), 40 types of concrete specimen were exposed to seawater at the Träslövsläge field site on the west coast of Sweden (Figure 5.31). The specimens were periodically sampled over 10 years and their chloride penetration profiles determined (Tang, 2003a). These field data serve as the 'first-hand' information about chloride ingress into concrete, and are valuable for the validation of prediction models of chloride penetration. Parts of the data have been used in the benchmarking evaluation in the EU CHLORTEST project, as discussed in Section 5.5.

Figure 5.31 The field exposure site at Träslövsläge harbour, on the west coast of Sweden.

Field exposure site and concrete specimens

The field exposure site is situated in Träslövsläge harbour in the south west of Sweden, 80 km south of Gothenburg. The field site is shown in Figure 5.31. Three pontoons are placed behind a pier in the harbour. The chloride concentration in the seawater varies from 10 to 18 g Cl per litre, with an average value of about 14 g Cl per litre. The typical water temperature is given in Figure 5.32; the annual average value is about +11°C.

At this field exposure site, 40 different types of concrete were tested. The main differences between the concretes were in the water/binder ratio (0.25, 0.3, 0.35, 0.4, 0.5, 0.6 to 0.75), the binder type (four types of cement with different additions of silica fume and fly ash), and air content (6% and 3% entrained air, and natural air without air-entraining). Concrete slabs of $1000 \times 700 \times 100$ mm were cast in the laboratory. After about 2 weeks of moisture curing, the slabs were transported to the field site and mounted on the sides of the pontoons for exposure, with the bottom sides of the slabs facing the seawater. Parallel slabs called 'null' ('zero') specimens were used for the measurement of chloride transport properties. A detailed description of the concretes and the measurement techniques can be found in the report by Tang (2003b).

Models considered for the validation

MODEL 1: A SIMPLE ERFC MODEL

Model 1 is the simplest error function solution to Fick's second law – an erfc model with constant D and C_s (see Section 4.4.2). The value of D_{RCM},

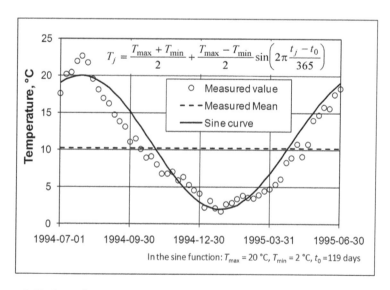

Figure 5.32 Annual seawater temperature at Träslövsläge harbour.

measured in the laboratory at a concrete age of 6 months using the RCM test (see Section 3.3.4), was used as the constant D, and the C_s value was calculated using the following equation:

$$C_s = A_{s, Cl}(w/b)$$ (5.37)

where $A_{s, Cl}$ is a regression parameter describing the relationship between the chloride surface content and the water/binder ratio (Engelund *et al.*, 2000).

MODEL 2: A MODIFIED ERFC MODEL

Model 2 is a modified erfc model with a time-dependent apparent diffusion coefficient D_a and constant C_s (see Section 4.4.3). The time-dependent D_a is expressed in one of two ways:

- D_a may be expressed as proposed by DuraCrete (Engelund *et al.*, 2000), i.e.

$$D_a = k_e k_c D_{RCM} \left(\frac{t_0}{t}\right)^n$$ (5.38)

where k_e is the environmental factor, k_c is a factor that accounts for the influence of curing on D_{RCM}, t_0 is the concrete age at which D_{RCM} is measured, t is the exposure duration, and n is the age factor describing the time dependence of the effective diffusion coefficient. In the Dura-Crete model (Engelund *et al.*, 2000), the concrete age t_0 is specified as 28 days, while in the Swedish data D_{RCM} was measured at a concrete age of about 6 months. However, if the same n value is assumed for the period from 28 days to 6 months, there should be no difference in the D_a values calculated using $D_{RCM\ 28d}(t_{28d})^n$ and $D_{RCM\ 6m}(t_{6m})^n$.

- D_a may be expressed as follows:

$$D_a = \frac{k_e k_c D_{RCM}}{1-n} \left[\left(1+\frac{t_{ex}}{t}\right)^{1-n} - \left(\frac{t_{ex}}{t}\right)^{1-n}\right]\left(\frac{t_0}{t}\right)^n$$ (5.39)

which is mathematically equal to Eq. 4.23 if D_{test} in Eq. 4.23 is replaced by $(k_e k_c D_{RCM})$ and the exposure duration $(t-t_{ex})$ in Eq. 4.23, where t is defined as concrete age, is replaced by t, where t is defined as the exposure duration.

The same n values were used in both alternatives, although mathematically the age factor n is different in Eqs 5.38 and 5.39, as clarified in Eq. 4.24. The purpose here in the validation is to examine the closeness of the predictions obtained with n values reported in the literature and the field data.

MODEL 3: THE CLINCONC MODEL, ENGINEERING EXPRESSION

Model 3 is a mechanism-based model, called ClinConc, which predicts the free chloride penetration through the pore solution in concrete using a flux equation based on the principle of Fick's law, and then converts the free chloride concentration to the total chloride content (see Section 4.5.3). The model includes a series of equations, but the calculations can be simply done in a worksheet. In recent years, this model has been developed for more engineer-friendly applications (Tang, 2007, 2008). In this engineering expression of ClinConc, the results of original numerical iterations for the calculation of the free chloride transport process are summarised using a single error-function equation:

$$\frac{c - c_i}{c_s - c_i} = 1 - \mathrm{erf}\left(\frac{x}{2\sqrt{\frac{\xi_D D_{RCM\,6m}}{1 - n'}\left(\frac{t_{6m}}{t}\right)^{n'}\left[\left(1 + \frac{t_{ex}}{t}\right)^{1-n'} - \left(\frac{t_{ex}}{t}\right)^{1-n'}\right]t}}\right) \tag{5.40}$$

where c, c_s and c_i are the concentration of free chloride ions in the pore solution at depth x, at the surface of the concrete and initially in the concrete, respectively, n' is the age factor accounting for the decrease in diffusivity with exposure time t due to the time-dependent chloride binding, and ξ_D is the factor bridging the laboratory-measured $D_{RCM\,6m}$ and the initial apparent diffusion coefficient for the actual exposure environment:

$$\xi_D = \frac{\left(0.8a_t^2 - 2a_t + 2.5\right)\left(1 + 0.59K_{b\,6m}\right)e^{\frac{E_D}{R}\left(\frac{1}{293} - \frac{1}{T}\right)}}{1 + k_{OH\,6m}K_{b\,6m}f_b\beta_b\left(\frac{c_s}{35.45}\right)^{\beta_b - 1}e^{\frac{E_b}{R}\left(\frac{1}{T} - \frac{1}{293}\right)}}k_D \tag{5.41}$$

where a_t is a factor that describes how the chloride binding changes over time, f_b and β_b are chloride binding constants, E_D and E_b are the activation energy of chloride diffusion and binding, respectively, k_D is the expansion factor (which depends on the type of binder and the water/binder ratio), and $k_{OH\,6m}$ and $K_{b\,6m}$ are factors accounting for the effects of hydroxyl ion concentration in the pore solution, gel content and water-accessible porosity at the age t_{6m}:

$$k_{OH6m} = e^{0.59\left(1 - \frac{0.043}{[OH]_{6m}}\right)} \tag{5.42}$$

and

$$K_{b6m} = \frac{W_{gel6m}}{1000}\,\varepsilon_{6m} \tag{5.43}$$

where $[OH]_{6m}$, $W_{gel\ 6m}$ and ε_{6m} are the hydroxide concentration (mol/$m^3_{pore-solution}$), gel content (kg/$m^3_{concrete}$) and water-accessible porosity at the age t_{6m}.

According to the data obtained from the 10-years exposure in the seawater at Träslövsläge (Tang, 2003a), the expansion factor k_D can be estimated as:

$$k_D = \begin{cases} 1+8\left[0.4-(w/b)\right]+7SF+ \\ \qquad 3800\left(SF \times FA\right)(SF+FA) & \text{for } 0.25 \leq w/b \leq 0.4 \\ 1 & \text{for } w/b > 0.4 \end{cases} \qquad (5.44)$$

where w/b is the water/binder ratio, and SF and FA are the mass fractions of silica fume and fly ash to the total binder content, respectively. This expansion factor describes the ratio of the diffusion coefficient measured in the field to that measured in the laboratory test.

Similarly, the binding factor a_t can be estimated as:

$$a_t = \begin{cases} 0.36+1.4\left[0.4-(w/b)\right]+0.4SF+ \\ \qquad 38\left(SF \times FA\right)(SF+FA) & \text{for } 0.25 \leq w/b \leq 0.4 \\ 0.36+1.4\left[0.4-(w/b)\right] & \text{for } 0.4 < w/b \leq 0.6 \end{cases} \qquad (5.45)$$

In Eqs 5.44 and 5.45 the effects of pozzolanic additions become insignificant when the water/cement ratio is > 0.4, probably because of the sufficient volume of the capillary network in a concrete with a sufficiently high water/cement ratio (> 0.4). This capillary volume can release the stresses that occur due to post-hydration pozzolanic additions, and the capillary network provides sufficient pathways for the gel to bind with the penetrated free chloride ions. Owing to the complicated preparation of the test samples used for the measurement of chloride binding, the available values for the chloride binding constants f_b and β_b are very limited. Based on the data reported by Tang (1996b) and the practical application of these data in later modelling work (Tang, 2003a), the values $f_b = 3.6$ and $\beta_b = 0.38$ are quite applicable to concrete containing Portland cement or silica fume. The following equations may be used to estimate these constants for binders containing fly ash

$$f_b = 3.6+7FA \quad \text{and} \quad \beta_b = 0.38-0.3FA \qquad (5.46)$$

and slag

$$f_b = 3.6+3.5BFS \quad \text{and} \quad \beta_b = 0.38-0.14BFS \qquad (5.47)$$

where *FA* and *BSF* are the mass fractions of fly ash and slag, respectively, with respect to the total binder.

The age factor n' is, according to the numerical simulation done by Tang (2008), mainly attributed to the increase in chloride binding through the parameter a_t, and can be expressed as the following regression equation:

$$n' = -0.45a_t^2 + 0.66a_t + 0.02 \qquad (5.48)$$

It should be noted that the above equation is based on the submerged environment, where the chloride solution is constantly in contact with the concrete surface. When the model is used for the atmospheric zone or a road environment, certain modifications are needed, as discussed in the following sections.

The total chloride content is the sum of the free and bound chloride ions:

$$C = \frac{\varepsilon(c_b + c)}{B_c} \times 100 \text{ mass\% of binder} \qquad (5.49)$$

where ε is the water-accessible porosity at the age after the exposure, B_c is the cementitious binder content ($kg/m^3_{concrete}$), and the bound chloride is calculated as:

$$c_b = f_t k_{OH} K_{b\,6m} f_b c^{\beta_b} e^{\frac{E_b}{R}\left(\frac{1}{T} - \frac{1}{293}\right)} \qquad \text{g/l} \qquad (5.50)$$

where f_t is a factor accounting for the time dependence of chloride binding. This factor can be calculated as:

$$f_t = a_t \ln\left(\frac{c - c_i}{c_s - c_i} t + 0.5\right) + 1 \qquad (5.51)$$

The parameter k_{OH} describes the effect of hydroxyl ions in the pore solution. If there is no leaching or carbonation of hydroxyl ions, k_{OH} will be equal to $k_{OH\,6m}$, as originally given by Tang (2007). If the leaching or carbonation of hydroxyl ions is taken into account, k_{OH} will be modified, as will be seen in Section 5.6.2. The other parameters, such as the porosity, gel content and hydroxyl ion concentration, are dependent on individual mix proportions, and can be estimated using well-established equations available in concrete handbooks or using the equations collected in Tang (1996b).

Input parameters

Some typical concrete mixes and relevant input parameters for different models are listed in Table 5.18, where the values of D_{RCM} were measured

Table 5.18 Typical concretes exposed at the Träslövsläge field site, Sweden, and input parameters used for modelling

Mix No.	Binder type	Binder content (kg/m³)	Water/ binder ratio	Air (vol % of concrete)	28-day compressive strength (MPa)	$D_{RCM\,6m}$ (×10⁻¹ m²/s)	Models 1 and 2		Model 3		
							C_s (% of binder)	n	$W_{gel\,6m}$ (kg/m³)	ε_{6m} (m³/m³)	$[OH]_{6m}$ (mol/l)
1	SRPC	490	0.30	3.7	96	2.52	3.75	0.3	323	0.102	0.652
2		450	0.35	6.0	70	3.6	4.38		356	0.103	0.582
3		420	0.40	6.2	58	12.2	4.12		395	0.107	0.521
4	10% SF	500	0.30	1.1	117	0.34	3.75	0.62	310	0.116	0.514
5	20% FA	630	0.30	3.1	98	1.49	3.24	0.69	336	0.119	0.563
6	5% SF + 10% FA	450	0.35	6.4	73	1.04	4.38	0.69	326	0.104	0.489
7	5% SF	550	0.25	1.3	125	0.85	3.13	0.62	269	0.104	0.674
8		0.30	0.8	112	0.62	3.75			314	0.109	0.581
9		0.35	5.8	72	2.93	4.38			351	0.110	0.520
10		0.40	6.1	61	4.43	5.0			389	0.115	0.462

Note
FA, fly ash; SF, silica fume; SRPC, sulphate-resistant Portland cement.

at a concrete age of 6 months, the C_s and n values (for Models 1 and 2) were calculated using Eq. 5.38 in such a way as described in the Duracrete, cf. Engelund *et al.*(2000) and the n values (for Model 2 only) were taken from the Duracrete, cf. Engelund *et al.*(2000). The values of W_{gel6m}, ε_{6m} and $[OH]_{6m}$ (for Model 3 only) were estimated based on the binder type and the conventional equations for degree of hydration, as summarised in Tang (1996b).

For Model 2, a curing factor $k_c = 0.85$ was assumed for $t_{ex} = 14$ days and the environmental factor $k_e = 1.32$ was taken for the submerged zone, according to the DuraCrete (Engelund *et al.*, 2000).

For Model 3, the free chloride concentration $c_s = 14$ g/l and the annual average temperature $T = 11°C$, as measured at the Träslövsläge site, were used. The activation energies $E_b = 40\,000$ J/mol and $E_D = 42\,000$ J/mol, as previously used in Tang and Nilsson (1996), were also used. Other parameters were calculated using Eqs 5.41 to 5.51.

Comparison between modelled and measured results

The measured data after 10 years of exposure and the modelled profiles for different types of concrete are shown in Figures 5.33 and 5.34. The results show that, for Portland cement concrete (Figure 5.33, Mixes 1–3), Models 2a and 3 predict the chloride profiles relatively closely to the measured ones, while Models 1 and 2b overestimate the chloride ingress. For the concrete with blended cements (Mixes 4–10 in Figures 5.33 and 5.34), Models 2b and 3 give better predictions, while Model 2a in most cases underestimates the chloride ingress. As expected, in general, Model 1 overestimates the chloride ingress.

It is expected that Model 3 will give good predictions, because this model was developed using the Träslövsläge field data and describes chloride ingress in a relatively scientific way. In the model, only free chloride ions can diffuse, and chloride binding is described by a non-linear binding isotherm that is a function of pH, temperature and the duration of chloride contamination. Therefore, the actual environmental conditions can easily be used as the boundary conditions in the modelling.

It is obvious that Model 1 would overestimate chloride ingress, because this model does not take into account the effect of chloride binding, which can significantly retard chloride ingress into concrete.

Model 2a in most cases underestimates the chloride ingress. The underestimation is, to a large extent, related to the magnitude of the age factor n. The larger the n value, the greater the underestimation. Mixes 5 and 6 have highest n value (0.69) in this study, and consequently show the largest underestimation with this model, as can be seen in Figure 5.33.

Model 2b accounts mathematically for the instantaneous time-dependent diffusion process. However, it still does not take into account the effect of chloride binding. Therefore, it is mere coincidence that this

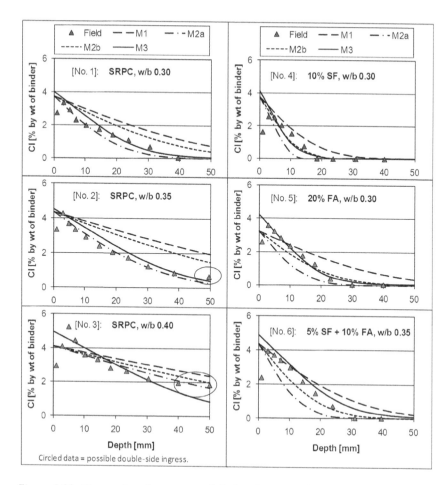

Figure 5.33 Comparison between modelled and measured chloride profiles after 10 years of exposure in seawater at Träslövsläge: (left) SRPC concrete; (right) fly ash concrete.

model gives good predictions like those obtained using Model 3, although in some case the chloride ingress overestimated (see Figure 5.33, Mixes 1–3).

Figure 5.35 shows some examples of the predicted chloride ingress after 100 years of exposure in seawater. Assuming a 1% chloride content as a threshold, Model 1 gives a chloride penetration depth almost twice that given by Model 3 gives, while the prediction for 100 years of exposure given by Model 2 is close to the actual chloride ingress at 10 years as measured at the field exposure site. This indicates that attention must be paid to the selection of proper *n* values when using modified erfc models.

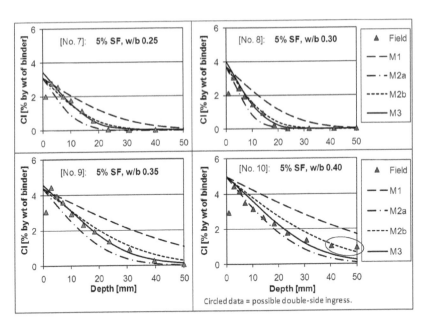

Figure 5.34 Comparison between modelled and measured chloride profiles after 10 years of exposure in seawater at Träslövsläge: silica fume concrete.

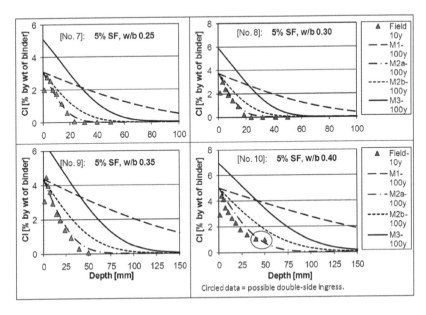

Figure 5.35 Predicted chloride profiles after 100 years of exposure in seawater at Träslövsläge: silica fume concrete.

5.6.2 *Data collected over 10 years of exposure in a road environment*

In the mid-1990s, as a successor to project BMB (Sandberg, 1996), a new Swedish national project BTB, 'Durability of Concrete Structures in a De-icing Salt Environment' was initiated. Since 1996, a number of reinforced concrete specimens with different qualities have been exposed at a site next to Highway 40 between Borås and Gothenburg, where de-icing salt is used extensively on the road due to the severe winter climate. The chloride profiles in some of the concrete specimens were measured after 1, 2, 5 and 10 years of exposure. These field data were valuable in the validation of prediction models presented here.

Field exposure site at Highway 40

The field exposure site at Highway 40 was established in the autumn of 1996. The site comprised a gravel area (200 m long and a couple of metres wide) next to the highway, and the concrete specimens were mounted in steel frames at road level, as shown in Figure 5.36. A guard rail was installed to separate the site from the traffic. The rail was placed in such a way as to ensure traffic safety while fully exposing the specimens to the splash water from passing traffic. The climate at the site was moist, and the specimens

Figure 5.36 The field exposure site at Highway 40: (top) schematic drawing of the site; (bottom) the exposure site (left) and the specimens placed in steel frames behind a guard rail (right).

were exposed during the winter to low temperatures and de-icing salts, producing a climate corresponding to exposure class XD 3/XF 4 in DIN EN 206-1:2001.

Highway 40 leads from Gothenburg to the east, through Borås and towards Jönköping. Over the year, the daily average number of vehicles passing the field site is around 12,000, of which 1250 are heavy vehicles (data from measurements carried out by the Swedish Road Administration in 2000).

For safety reasons, de-icing salts are used during the winter in many parts of Sweden to keep road surfaces free from snow and ice. The de-icing agent used is sodium chloride, which is spread either in the form of a solution (about 24% NaCl), as a preventive measure, or as crystals when spread on snow. In this region, de-icing salts are normally used between October and April. Table 5.19 shows the estimated total amount of salt spread on the highway per square metre per year. The figures in Table 5.19 are based on the data from the Swedish Road Administration. The table also shows the number of occasions on which de-icing salts were spread on the highway in each winter season. As can be seen from Table 5.19, the amount of salt spread on the road was markedly reduced around year 2000. This was due to the introduction of a new method of applying salt, as a solution, which uses a smaller amount of salt.

The annual precipitation in this region varies from 800 mm to 1800 mm, with an average of 1170 mm according to the information provided by the Swedish Meteorological and Hydrological Institute (SMHI).

The monthly air temperature registered at the climate station near the field exposure site is shown in Figure 5.37. From these data an annual average temperature of 7°C was estimated. If the freezing period (assuming that at these temperatures there would be a cessation in the diffusion process) is excluded, the average temperature would be about 10°C.

Table 5.19 Estimated total annual amount of salt spread on Highway 40 and the number of occasions on which de-icing salts were spread on the road in each winter season

Winter	*1996–97*	*1997–98*	*1998–99*	*1999–2000*	*2000–01*	*2001–02*
Amount of salt (kg/m²)	1.9	2.4	2.3	2.1	1.1	1.2
Number of occasions	126	157	151	141	117	148

Winter	*2002–03*	*2003–04*	*2004–05*	*2005–06*	*2006–07*	
Amount of salt (kg/m²)	1.1[1]	1.3[1]	1.3[1]	1.2[1]	1.1[1]	
Number of occasions	128	156	151	141	123	

Note
1 Estimated from the number of occasions.

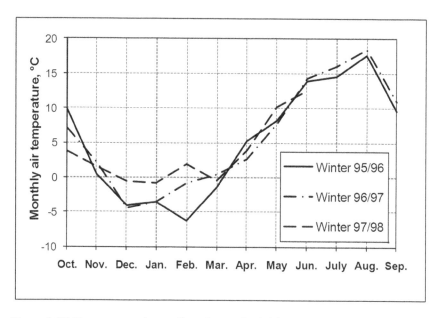

Figure 5.37 Temperature data collected near the field exposure site at Highway 40.

It should be noted that the actual chloride concentration in the highway environment is unknown, although some data of salt spread are available (see Table 5.19). Wirje and Offrell (1996) investigated chloride ingress into mortar specimens placed at different locations along the road. The results showed that the ingress of chloride decreases with increasing height above road level. Results presented by Tang and Utgenannt (2000) from splash water collected at different locations around the exposure site confirm the findings of Wirje and Offrell (1996).

Concrete specimens

Similar to the BMB project, over 30 different types of concrete were tested at this field exposure site. The main differences between the concretes include water/binder ratios of 0.3–0.75, binder types with different additions of silica fume, fly ash, blast-furnace slag and limestone filler, and air entraining (mainly for frost resistance). Two types of concrete specimen, one a plain concrete block with dimensions 400 × 300 × 300 mm and the other a reinforced concrete beam with dimensions 1200 × 30 0 × 300 mm, were cast and cured in the laboratory for 35–70 days before being placed at the field site. The plain concrete blocks were designed for the sampling of chloride penetration profiles, and the reinforced concrete beams were designed for the testing of corrosion resistance under uncracked and pre-cracked conditions. Detailed information about the concrete specimens and the mix proportions is given in Tang and Utgenannt (2007).

Models and input data

Two models, the DuraCrete and ClinConc models, were evaluated using the data obtained from the Highway 40 field exposure site. In this evaluation, two concrete mixes, similar to Mix 3 (SRPC binder, water/binder ratio 0.4) and Mix 10 (5% silica fume, water/binder ratio 0.4) those in Table 5.18, were considered.

For the DuraCrete model (Engelund *et al.*, 2000), there were no input parameter values available for a de-icing salt road environment. Although some values were suggested by the Swedish Association of Concrete (Betongföreningen, 2007), it has been demonstrated that these values result in a significant underestimation of chloride ingress (Tang and Utgenannt, 2007). Therefore, values for an atmospheric zone were used in this evaluation, as listed in Table 5.20. An initial chloride content $C_i = 0.02\%$ of binder was assumed in the modelling.

The ClinConc model described in Section 5.6.1 was based on the submerged marine environment, where the chloride solution is constantly in contact with the concrete surface, and no carbonation or leaching of hydroxyl ions was taken into account. When the model is used for a road environment, certain modifications are needed.

First, the surface free chloride concentration in the de-icing salt highway environment is unknown, and can vary from zero to near saturation in a short period of time. Second, the time-dependent chloride binding due to further hydration and pore saturation under the marine environment may not occur to the same extent in the de-icing salt road environment. Thirdly, the existing CO_2 in the road environment will react to some degree with the alkaline components of the concrete, resulting in a decreased pH of the pore solution. This will, in turn, reduce the chloride binding (although this decreased pH value cannot be detected using phenolphthalein). There is a lack of available data for quantifying the above influencing factors, especially the third one (leaching or partial carbonation of hydroxyl ions, and the effect of this on chloride binding). Therefore, In this study, after some

Table 5.20 Input parameters for the DuraCrete model applied to the road environment

Binder type	100% SRPC	95% SRPC + 5% SF
Curing factor, k_c	0.79	
Environmental factor, k_e	0.68	
$A_{s,\,Cl}$ (mass% of binder)	2.57	3.23
Age factor, n	0.65	0.79

preliminary trials using chloride-profile data obtained from the field site, the following two modifications were made:

1   the time-dependent factor of chloride binding a_t was taken as one-third that for the submerged marine environment (see Eq. 5.45);
2   a factor k_{carb}, which takes into account the leaching or partial carbonation of hydroxyl ions, was added to the parameter k_{OH} in Eq. 5.50:

$$k_{OH} = e^{0.59\left(1-\frac{0.043}{k_{carb}[OH]_{6m}}\right)} \tag{5.52}$$

As a preliminary consideration, the factor k_{carb} can be expressed as

$$k_{carb} = \begin{cases} A_{carb}\left(\dfrac{0.4}{w/b}\right)^2 & \text{for } 0.25 \le w/b \le 0.4 \\[2mm] A_{carb} & \text{for } w/b > 0.4 \end{cases} \tag{5.53}$$

where $A_{carb} = 0.1$ for SRPC and 0.2 for (95% SRPC + 5% SF).

The value of $A_{carb} = 0.1$ implies a reduction in the hydroxyl ion concentration to 1/10 of the initial one, corresponding to a reduction in the pH value of 1. For Portland cement concrete, the initial pH value in the pore solution is about 13.6. After a certain degree of leaching of hydroxyl ions, the pH value reduces to 12.6, which is the equilibrium concentration of hydroxyl ions in portlandite. For pozzolanic binder the initial pH value in the pore solution is lower than that in Portland cement, and thus it requires less reduction in the pH to reach the equilibrium of portlandite (pH 12.6). Therefore, the A_{carb} value for pozzolanic binder could be higher than 0.1. With further carbonation, the A_{carb} value can be even lower for lower pH values. This means that, due to leaching or carbonation of hydroxyl ions, the actual [OH-] concentration in the pore solution is reduced by the factor k_{carb}, resulting in a lower value of k_{OH}, or less chloride binding.

Obviously, the above modifications are empirical, and further numerical simulations are needed in order to clarify these modifications 'mechanistically'.

The environmental data used in the ClinConc model include the free chloride concentration $c_s = 1.5$ g/l and the annual average temperature $T = 10°C$ (excluding the freezing period). The initial free chloride concentration in the pore solution was assumed to be 0.002 g/l. The concrete age at the start of exposure t_{ex} was 28 days. Other parameters are similar to those used in Section 5.6.1.

Comparison between the predicted and the measured chloride profiles

The predicted and measured chloride profiles are shown in Figures 5.38 and 5.39. It can be seen that none of the models predicts the chloride ingress precisely, and for the silica fume concrete the profiles at 5 years are in some cases higher than those at 10 years. In these cases the DuraCrete model overestimates the chloride ingress in the SRPC concrete and underestimates that in the silica fume concrete. In addition, the predicted profiles at 1.5–10 years are very close each other, which does not really reflect the reality. The

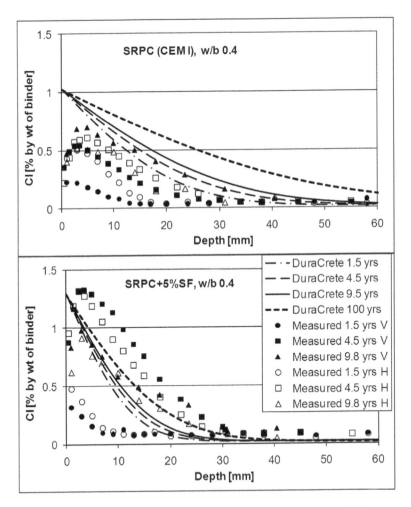

Figure 5.38 Profiles modelled using the DuraCrete model for the field exposure site at Highway 40. H denotes field data measured from the horizontal exposure surface, and V denotes the field data measured from the vertical exposure surface.

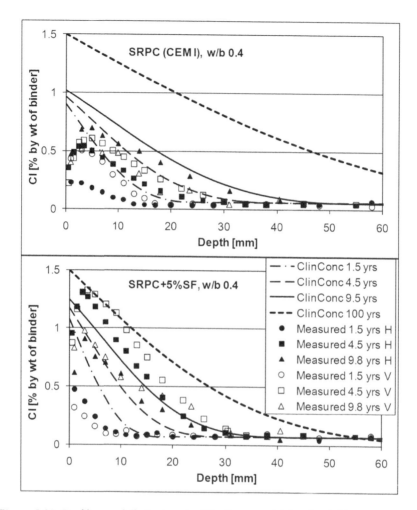

Figure 5.39 Profiles modelled using the ClinConc model for the field exposure site at Highway 40.

ClinConc model gives relatively good predictions, although the profile at 5 years for silica fume concrete is far from the measured ones. Measurement uncertainty in the field data could also be a reason for this unpredictability.

5.6.3 Data from real structures

Although there are some case studies reported in the literature of chloride ingress into old concrete structures, actual data of chloride profiles and other input parameters relevant to modelling are rarely available. In addition, time and resources are required to reach an understanding of published data. Therefore, only those data that are both available and understood by the authors were used for the validation reported in this section.

Chloride profiles from real road bridges

At the end of 1990s, a number of road bridges that were 25–30 years old and located around Gothenburg, Sweden, were sampled to determine their chloride-ingress profiles (Lindvall, 2001). Detailed information about the bridges and the samples collected are given in Lindvall (2001). Some of the profiles obtained for concrete elements near a heavy traffic lane were used in this validation of the DuraCrete and ClinConc models. The concrete used during the period when the bridges were constructed was produced with Swedish SRPC. According to measurements made using the RCM test on specimens taken from deeper parts of the cores, the D_{RCM} values are in the range 8.6–16.6×10^{-12} m²/s (Lindvall, 2001). Considering the concrete age (25–30 years), we can assume that the water/cement ratio w/c is in a range 0.4–0.5. Therefore, the mixes used in the BTB project (Tang and Utgenannt, 2007), with SRPC and w/c = 0.4 and 0.5, were used in the modelling, with $D_{RCM\ 6m} = 8.6 \times 10^{-12}$ m²/s for w/c = 0.4 and 16.6×10^{-12} m²/s for w/c = 0.5. Other input parameters are the same as those used in Section 5.6.2. The modelled results are presented in the following Figures 5.40, 5.41 and 5.43.

Figure 5.40 shows chloride profiles taken from two bridges, 'O 670' and 'O 707'. The profile marked 'O 670 KK +RH' was taken from the side

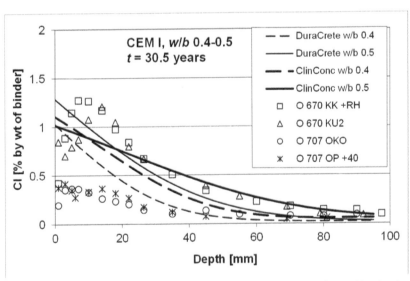

Figure 5.40 Comparison between modelled and measured chloride profiles for four 30-year-old road bridges.

beam, approximately 3 m from the traffic lane, of a highway bridge with heavy traffic at high speed (>100 km/h). The profile marked 'O 670 KU2' was taken from the underside of the pavement slab near the joint where the leakage of water was visible. The profile 'O 707 OKO' was taken from the side beam, about 0.5 m from the traffic lane, of a slip way with fairly little traffic at very low speed (20–30 km/h), and the profile 'O 707 OP +40' was taken from a column of the bridge, approximately 2.3 m from the traffic lane, of a local road with little traffic at low speed (30–40 km/h). It seems that the ClinConc model gives a fairly good prediction of the chloride ingress in concrete near the heavy traffic lane (Bridge 'O 670'), while the DuraCrete model slightly underestimates the chloride ingress in concrete near the heavy traffic lane.

The chloride profiles shown in Figure 5.41 are from samples taken from Bridge O 951 over the highway between Gothenburg and Malmö, which carries heavy traffic at high speed (> 100 km/h). The profiles were taken from the lower part of the first column (nearest to Malmö (denoted by U), approximately 2 m from the traffic lane. The sampling positions with reference to the direction of traffic flow are illustrated in Figure 5.42. The Clin-Conc model predicted three of the profiles fairly well, but not the highest one, which was taken from the first column on the side receiving splash water from vehicles travelling from Malmö. The DuraCrete model underestimates all four chloride profiles from this bridge column.

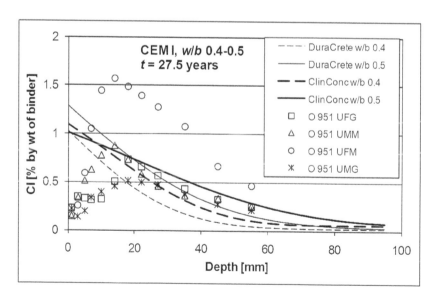

Figure 5.41 Comparison of modelled and measured chloride profiles for column U (the column nearest to Malmö) of bridge O 951 (27 years old) over Highway E6. FG, FM, MG and MM indicate the direction of traffic at the sampling positions (see Figure 5.42).

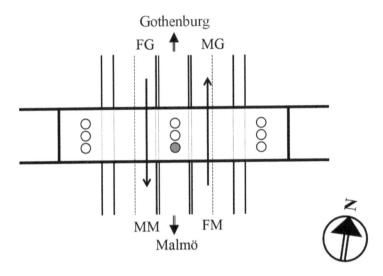

Figure 5.42 The sampling positions on column U of bridge O 951. The examined column is shaded in grey. (Based on Lindvall (2001).)

Figure 5.43 shows chloride profiles in samples taken from bridge O 978 over Highway 40 between Borås and Gothenburg, which carries heavy traffic at high speed (> 100 km/h). The profiles were taken from the lower part of the first column (the column nearest to Gothenburg), approximately 3 m from the traffic lane. In this case the ClinConc model overestimates and the DuraCrete model slightly overestimates the measured chloride profiles.

SUMMARY

It can be seen from the above cases that the ClinConc model gives, in general, fairly good predictions, the predicted profiles being close to the measured ones, while the DuraCrete model, with the currently available input parameters, seems in some cases to underestimate chloride ingress into concrete near heavy traffic lanes.

Chloride profiles from real marine structures

Tang and Hassanzadeh (2009) have made an investigation of the chloride ingress into an old marine concrete structure after 30 years of exposure in the Baltic Sea, where the chloride concentration is about 5 g/l. Actual chloride penetration profiles were measured in samples taken from different exposure zones of the structure. Cores taken from parts of the structure not contaminated with chloride were used in the determination of the chloride diffusion coefficient by the RCM test. The DuraCrete and the ClinConc models were

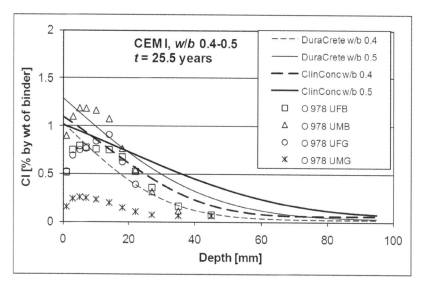

Figure 5.43 Comparison of modelled and measured chloride profiles for column U (the column nearest to Gothenburg) of bridge O 978 (25 years old) over Highway 40. The direction of traffic at the sampling positions: FB, from Borås, FG, from Gothenburg; MB, towards Borås; MG, towards Gothenburg.

used to predict chloride penetration profiles, which were then compared with the actual profiles. Detailed information on the concrete mix type and the modelling used are given in Tang and Hassanzadeh (2009).

The results predicted using the DuraCrete model and the recommended input parameters for this model (Engelund *et al.*, 2000) are shown in Figure 5.44, together with actual chloride penetration profiles were measured in six cores (numbered 307, 308, 314, 403, 404 and 405) taken from the submerged zone. It can be seen that the predicted curves are too high compared with the measured ones. Figure 5.45 shows the results obtained using one-third of the chloride concentration factor, i.e. $A_{s, Cl} = 3.4\%$ instead of 10.3% of binder, as recommended in the DuraCrete model (Engelund *et al.*, 2000). In this case the predicted chloride-ingress profiles are comparable with some of the measured profiles (cores 307, 403 and 404), but the model still cannot account for the profiles for cores 308, 314 and 405.

It can be seen from Figure 5.44 that the measured chloride profiles vary greatly, even though the samples were taken from the same submerged zone. An explanation for this is the leaching or partial carbonation of hydroxy ions, which has been proven by Hassanzadeh (2007). According to his measurements, the pH values found in cores from a submerged zone vary from 12.1 at depths of 0–5 mm to 12.6 at deeper depths. Therefore,

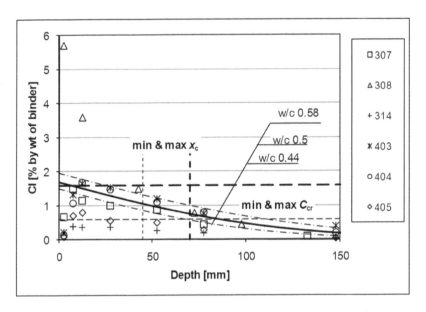

Figure 5.44 Results obtained using the DuraCrete model and the input data recommended for the model (Engelund *et al.*, 2000).

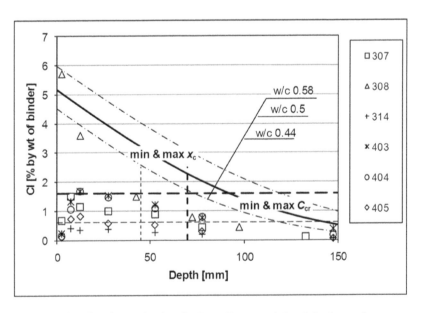

Figure 5.45 Results obtained using the DuraCrete model and the input data recommended for the model (Engelund *et al.*, 2000), but with $A_{s,Cl} = 3.4\%$ of binder.

when using the ClinConc model to predict the profiles, the effect of the pH value on chloride ingress was taken into account by changing the initial hydroxyl ion concentration, i.e. the value of $[OH]_{6m}$ in Eq. 5.42. The results are shown in Figure 5.46. Obviously, a decrease in the pH value dramatically reduces the amount of bound chloride, resulting in a flat chloride-ingress profile.

As a decrease in the value of $[OH]_{6m}$ results in a decrease in $k_{OK\,6m}$ or an increase in ξ_D (see Eq. 5.41), the apparent diffusion coefficient in Eq. 5.40 will also increase for the entire duration of exposure. This cannot be true, because leaching and partial carbonation are gradual processes that occur over the duration of exposure, starting from the surface zone of the concrete, and the diffusion front should not be influenced by this slow process. Therefore, adding the factor k_{carb} in Eq. 5.52 could be a better way to treat the effect of leaching or carbonation, because the k_{carb} value influences the amount of bound chloride only, which in turn influences the total chloride distribution in Eq. 5.49, but not the apparent diffusion coefficient in Eq. 5.40. As the initial pH of the pore solution is about 13.6, assuming $k_{carb} = 0.1$ and 0.035 corresponds to pH values of 12.6 and 12.1, respectively. The recalculated results are shown in Figure 5.47. In this way, the variation in the measured chloride profiles is better explained by the effects of leaching and partial carbonation of hydroxyl ions. The results

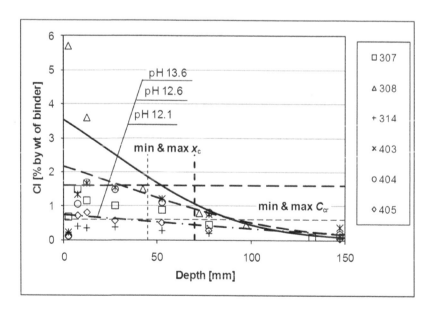

Figure 5.46 Results obtained using the ClinConc model and an initial hydroxyl ion concentration $[OH]_{6m}$ corresponding to pH 13.6, 12.6 and 12.1, respectively.

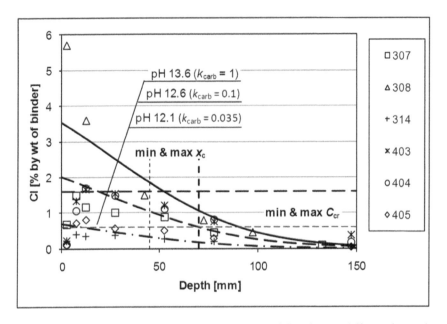

Figure 5.47 Results obtained using the ClinConc model and using different k_{carb} values corresponding to pH 13.6, 12.6 and 12.1, respectively.

also demonstrate that the ClinConc model can predict chloride ingress in real concrete structures.

Recently, the ClinConc model has been used to predict chloride ingress in a Portland cement concrete pier in the North Sea, near the Dornoch bridge, Scotland, after 18 years of exposure (Nanukuttan *et al.*, 2010). The predicted results are in good agreement with the measured profiles, although the model seems to slightly underestimate the real profile measured after 18 years of exposure, see Figure 5.48.

5.7 Conclusions

5.7.1 Conclusions on the sensitivity analysis of the models

The sensitivity analysis described in Sections 5.2 and 5.3 of how each model takes into account environmental differences, concrete characteristics and corrosion onset was dealt with theoretically, first using a probabilistic approach on the first 10 years of ingress for a selection of models, and then in a general way using erfc models.

The sensitivity analysis employing probabilistic methods was done using two input data: the surface chloride concentration and the chloride diffusion coefficient. This study has shown that the importance factors are dependent

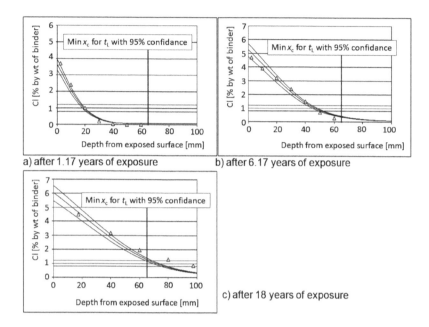

a) after 1.17 years of exposure   b) after 6.17 years of exposure

c) after 18 years of exposure

Figure 5.48 The real (points) and predicted (curves) chloride profiles after 1.17, 6.17 and 18 years of exposure in the North Sea. (Adapted from Nanu-kuttan *et al.* (2010).)

on time, and this was found with all three models tested, despite the fact that they are very different conceptually.

In the case considered, the influence of the surface concentration of chloride is more important than the influence of the diffusion coefficient, in all three models tested. Therefore, more effort should be spent on increasing the precision of measuring the surface concentration of chloride in order to increase the precision of predicted profiles. The surface concentration of chloride is a function of the environment, but also of the interactions between the cement paste and the chloride ions. Therefore, knowledge of the binding isotherm is as key to the prediction of chloride ingress as is knowledge of the diffusion coefficient.

The above conclusions are based on the sensitivity analysis done for exposure times of up to 10 years. The conclusions drawn are dependent on the exposure time, but for longer exposure times, such as 100 years, the conclusions may be different.

However, the long-term sensitivity analysis showed that the influence of C_s is constant, while the influence of D is dependent on C/C_s. In some empirical models the most sensitive parameter is the age factor.

5.7.2 Conclusions on the benchmarking models

The aim of the benchmarking evaluation described in Sections 5.4 and 5.5 was to evaluate the ability of the selected models to reproduce the real situation found in the field. Laboratory test results were used as input data to the models, and in-field performance data were used as the comparison.

It was found that there is a tremendous number of chloride-ingress models available, and most of them use laboratory test results as input data. will have One or several models will require the results obtained with any particular test method as input data. However, a very important conclusion is that most of the chloride-ingress models are still not very accurate in predicting reality.

The main drawback of the models is the description of the boundary conditions, i.e. the chloride content of the exposed surface in a certain environment. In particular, the increase in the chloride content over time is still not understood. Most models also require data obtained from field exposure tests to express the boundary conditions, and the results obtained using some test methods are not reliable.

The time dependence of the chloride diffusion coefficients is still not well understood, and is described in very different ways in different models. This time dependence is not the result of any test method.

The benchmarking results show that one model (Model 5, the ClinConc model), which was previously calibrated using most of the profiles from Series 1, gives reasonably good predictions, indicating the importance of calibrating models against reliable long-term data.

5.7.3 Conclusions on the validation against long-term exposure data

Nowadays more and more long-term field exposure data are available for the validation of prediction models, and model developers are encouraged to calibrate their models against these field data.

Empirical models either significantly overestimate (e.g. the simple erfc model) or, often, underestimate (e.g. the DuraCrete model) chloride ingress if the input parameters from short-term exposure are used for the prediction of long-term performance.

The ClinConc model was calibrated against the 10-year data obtained from both a marine and a road exposure site. The validation of this model against this field data obtained from real concrete structures after 18–30 years of exposure showed that this model gives fairly good predictions for the chloride ingress in the submerged zone of a marine environment under and also in a de-icing salt road environment with heavy traffic travelling at high speed, although some empirical parameters have to be assumed in the latter case. The key to the successful predictions obtained with the Clin-Conc model is the consideration within the model of the free chloride ion concentration as the driving force and of chloride binding as an interaction between chloride ions and hydrates.

Chloride binding is an important factor influencing both the ion transport process and the total chloride ion distribution. Further study is needed to clarify the time-dependent behaviour of the binding, and the effect of leaching or partial carbonation of hydroxyl ions.

6 Overall discussions and conclusions

6.1 Concretes in chloride environments

Concretes exposed to chloride-containing environments, such as seawater or roads where de-icing salts are used in very cold weather, may have durability problems because of the chloride-induced corrosion of the steel reinforcement. This premature deterioration of concrete structures has led to an increasing demand for methods that better predict this distresses, and that can be used to evaluate the suitability of concrete mixes to satisfy the desired service life.

As stated in Chapter 1, chloride ingress into concrete involves complex physical and chemical processes. The complexity arises from at least three sources, as identified in Chapter 1:

- The external environment is not constant. In marine environments the amount of chloride in contact with the concrete depends on whether the structure is fully submerged or located in the tidal zone, or only in contact with marine fog, while in road environments the intermittent use of de-icing salts in very cold weather makes it difficult to calculate the amount of chloride sprayed onto the concrete surface of a structure.
- Concrete is composed of different types of cement and binder, in different mix proportions, which means that the concrete in different structures may not be the same. Concrete is, therefore, not a single material, but many different ones, and the properties of each of its constituent materials evolve with age.
- The mechanism of chloride penetration is not confined to a single transport process (e.g. diffusion), but may be combined with convection (absorption), chemical and physical binding, and interactions with other coexisting ions. Changes in temperature, rain and sunshine introduce variations that should also be taken into account.

As chloride ingress in concrete plays an extremely important role in the durability of concrete structures, many methods have been proposed for evaluating this ingress. However, the complexity of the process has, until now, hindered general agreement on a single test method being reached. The methods can be divided into three categories: diffusion tests, migration tests and indirect tests based on resistivity or conductivity. Different inter-laboratory comparisons have been carried out to evaluate some of the test methods, including the immersion test (NT BUILD 433), the rapid chloride migration test (NT BUILD 492 as a non-steady-state migration test), the steady-state migration tests and the resistivity test. Based on the evaluation of test results obtained in the EU project CHLORTEST, three methods – the immersion test (NT BUILD 433), the rapid chloride migration test (NT BUILD 492) and the resistivity test – are now recommended.

6.2 Summary of frequently used test methods

In this section, the most frequently used test methods are summarised. It should be noted that some of these tests have certain limitations, as identified in Chapter 3. Section 6.3 lists those tests that have been recommended for use based on the CHLORTEST inter-laboratory comparison study.

6.2.1 The immersion test (NT BUILD 443)

This method is based on natural diffusion under a very high concentration gradient. The test gives values of D_{nssd} (the non-steady-state diffusion coefficient) and C_s (the surface total chloride content) by curve-fitting the measured chloride profile to an error-function solution of Fick's second law. From the values of D_{ns} and C_s, the parameter K_{Cr} (the penetration parameter) can be derived. The test is relatively laborious and takes a relatively long time (more than 35 days) to complete.

Specimens: Three specimens of diameter ≥ 75 mm and length ≥ 60 mm, with the cut surface as the test surface and an epoxy coating on all non-exposed surfaces.

Pre-conditioning: Specimens are immersed in saturated limewater until a constant weight is reached (normally a few days).

Test method: Specimens are immersed in a solution of 165 g NaCl per litre for at least 35 days, and then samples are obtained by grinding the specimen successively from the exposed surface. The chloride-penetration profile is measured by means of potentiometric titration to determine the total chloride content of each powder sample.

Test duration: At least 35 days for immersion, plus a few days for pre-conditioning, and a few days more for grinding and analysing chloride profiles.

Test results: Chloride-penetration profiles, the curve-fitted parameters D_{nssd} and C_s, and the derived parameter K_{Cr}.

Precision: Average repeatability COV 20%, 18% and 9% for parameters D_{nssd}, C_s and K_{Cr}, respectively, and reproducibility COV 28%, 22% and 14% for parameters D_{nssd}, C_s and K_{Cr}, respectively, according to the final evaluation results from the CHLORTEST project (Tang, 2005), noting that K_{Cr} is proportional to the square root of D_{nssd}, implying that the deviation of K_{Cr} is theoretically half of that of D_{nssd}.

6.2.2 *The rapid chloride permeability test (RCPT)*

This is the first rapid test using an external electrical field to accelerate the procedure. It is widely used in North America and some other regions such as the Middle East, although the test results (the charge passed, given as Coulomb values) are not directly related to the chloride diffusion coefficient.

Specimens: Three specimens of diameter 95–102 mm (3.75–4 inches) and thickness 50 mm, with the cut surface as the test surface.

Pre-conditioning: Vacuum saturation with de-aerated tap water (about 24 hours).

Test method: A 60 V DC external potential is imposed across the specimen, with the test surface exposed to a 3% NaCl solution (upstream cell) and the opposite surface to a 3 N NaOH solution (downstream cell). The current passing through the specimen is recorded at intervals of ≤ 30 minutes.

Test duration: 6 hours for testing, plus 24 hours for pre-conditioning.

Test results: Coulomb values as an indicator of chloride penetration.

Precision: Average repeatability COV 12% and reproducibility COV 18%, according to ASTM C1202-05.

6.2.3 *The rapid chloride migration (RCM) test (NT BUILD 492)*

This is a non-steady-state migration test using an external electric field to accelerate the chloride penetration. The test gives values of D_{nssm} (the non-steady-state migration coefficient). The test is relatively simple and rapid, with a test duration in most cases of 24 hours.

Specimens: Three specimens of diameter 100 mm and thickness 50 mm, with the cut surface as the test surface.

Pre-conditioning: Vacuum saturation with saturated limewater (about 24 hours).

Test method: A 10–60 V DC external potential is imposed across the specimen, with the test surface exposed to a 10% NaCl solution and the opposite surface to a 0.3 M NaOH solution for a known amount of time (in most cases 24 hours). The specimen is then split open, and the depth of penetration of the chloride ions measured using a colorimetric method.

Test duration: 6–96 hours (in most cases 24 hours), depending on the quality of concrete, for migration, plus 24 hours for pre-conditioning.

Test results: Non-steady-state migration coefficient D_{nssm} calculated using the average penetration depth of the chloride ions.

Precision: Average repeatability COV 15% and reproducibility COV 24% for the parameter D_{nssm}, according to the final evaluation results from the CHLORTEST project (Tang, 2005).

6.2.4 The steady-state migration test

In this test the chloride flux is determined by measuring the change in conductivity in the downstream solution. The test gives values of D_{ssm}, from the flux, and D_{nssm}, from the time lag. The test is relatively simple as the of chloride concentration is measured indirectly through a simple conductivity measurement. However, the reproducibility of this test method is not satisfactory.

Specimens: Three specimens of diameter 100 mm and thickness 20 mm, with the cut surface as the test surface.

Pre-conditioning: Vacuum saturation with demineralised water (about 24 hours).

Test method: A 12 V DC external potential is imposed across the specimen, with the test surface exposed to a 1 M NaCl solution (upstream cell) and the opposite surface to demineralised water (downstream cell). The conductivity in the downstream solution is measured at time intervals, and converted to a chloride concentration, until a constant increase in conductivity is reached.

Test duration: A few days, up to about 2 weeks (depending on the quality of the concrete) for migration, plus 24 hours for pre-conditioning.

Test results: The steady-state migration coefficient D_s calculated from the slope of the constant portion of the concentration–time curve (the constant flux), and the non-steady-state migration coefficient D_{ns} calculated from the intersection on the time axis of the constant portion of the concentration–time curve (time lag).

Precision: Average repeatability and reproducibility COV 22% and 76%, respectively, for the parameter D_s, according to the final evaluation results from the CHLORTEST project (Tang, 2005); and average repeatability and reproducibility COV 24% and 45%, respectively, for the parameter D_{ns} according to the pre-evaluation results from the CHLORTEST project (Castellote and Andrade, 2005).

6.2.5 *The resistivity test*

The resistivity test is an indirect measurement of the transport property of concrete, because the electrical resistance of concrete is related to the pore structure and the ionic strength of the pore solution.

Specimens: Three specimens of diameter 100 mm and thickness 50 mm, with the cut surface as the test surface.

Pre-conditioning: Vacuum saturation with distilled or demineralised water (about 24 hours).

Test method: A constant AC current is imposed across the specimen and the change in potential measured. The potential value is then used to calculate the resistance using Ohm's law.

Test duration: A few seconds or minutes for the resistivity measurement, plus 24 hours for pre-conditioning.

Test results: Resistivity ρ.

Precision: Average repeatability COV 11% and reproducibility COV 25% for the resistivity ρ, according to the final evaluation results from the CHLORTEST project (Tang, 2005).

6.2.6 **In situ** *test methods*

A few *in situ* methods are available for testing the resistance of concrete to chloride ingress. A single-point measurement of the chloride content is useful for assessing the condition of a structure, or the reinforcement if the measurement is done at the level of the reinforcement. Multi-point measurements of the chloride content yield more information regarding the rate of chloride ingress into the concrete. In general, both single- and multi-point measurements can only be performed in a structure that has been exposed to the environment for several years or after significant penetration of chloride ions has occurred. Therefore, accelerated chloride migration tests are preferred for the testing of both new and existing structures.

The *in situ* rapid chloride migration (RCM) test is a useful tool for examining the quality of concrete in existing structures, but this method is partly destructive due to the need for cutting a core from the structure at the test location. The *in situ* migration coefficient obtained from the Permit ion migration test has been found to correlate well with coefficients obtained from other laboratory-based diffusion and migration test methods. The only limitation of the Permit ion migration test is that it involves the use of chloride solutions, and thus the test area becomes contaminated with chloride ions.

Testing with embedded sensors, which measure electrical resistance, has several advantages, including the possibility to continuously monitor a structure. However, a limitation of this type of measurement is that it is difficult to distinguish between the effects due to water and chloride ingress and those due to chemical reactions in the concrete.

The Wenner four-probe surface resistivity test is a simple and effective tool for assessing the condition of cover concrete. This test has similar limitations to those encountered when using embedded electrical sensors.

No data are available for the precision of *in situ* test methods, due to the practical difficulties of undertaking an inter-laboratory comparison study for these methods.

6.3 Recommended test methods

Based on the EU project CHLORTEST the following three methods are recommended for testing chloride ingress in concrete:

1 the immersion test (based on NT BUILD 443) for the determination of the non-steady-state diffusion coefficient D_{nssd} and surface total chloride content C_s;
2 the rapid chloride migration test (based on NT BUILD 492) for the determination of the non-steady-state migration coefficient D_{nssm} under the standardised laboratory exposure condition;
3 the resistivity test (based on the version used in the CHLORTEST project (Tang, 2005)) for the determination of resistivity ρ as an indirect measurement of the transport property of concrete.

These three recommended test methods have a precision in an acceptable range, i.e. repeatability COV$\leq$20% (11–20%) and reproducibility COV$\leq$30% (24–28%), and therefore they are suitable for use in data comparisons and industrial applications.

The recommended test methods, with certain revisions and modifications, are described in the Appendix.

6.4 Interpretation of the test results

6.4.1 Results of the immersion test (NT BUILD 443)

The immersion test provides corresponding values of D_{nssd} and C_s, obtained by curve-fitting the measured chloride profile to an error-function solution of Fick's second law, under the assumption of constant chloride-binding capacity. In reality, the chloride-binding capacity is non-linearly dependent on the free chloride concentration, and varies with the type of cementitious binder, as has been reported by many researchers (Tritthart, 1989a,b; Byfors, 1990; Tang and Nilsson, 1993b). The total chloride content is the sum of the free chloride ions in the pore solution and the bound chloride ions on the surface of the hydrated cement. Therefore, the C_s value is dependent on the porosity and the type of binder. Even under the same exposure conditions, i.e. in solutions of the same chloride concentration, different types of binder will give different values of C_s.

As D_{nssd} is coupled with C_s in the curve-fitting, the value of D_{nssd} alone may not reflect the actual resistance of the concrete to chloride ingress. To properly interpret the test results, the value of both D_{nssd} and C_s should be taken into account. The penetration parameter K_{Cr} combines the effects of D_{nssd} and C_s and, therefore, better facilitates comparison of the results. An example is presented in Figure 6.1, where Mix A has a lower value of D_{nssd} and a higher value of C_s than Mix B, but both mixes have the same value of K_{Cr}.

It should be noted that the parameter K_{Cr}, which has the dimensions mm/√year, is mainly useful for comparison purposes, and does not necessarily indicate the actual penetration depth per square root of the year.

6.4.2 Results of the rapid chloride migration (RCM) test

The RCM test provides a value of D_{nssm}, under the assumption of constant chloride-binding capacity during the test. This assumption may hold better in the RCM test than in the immersion test (see Section 6.4.1), owing to the

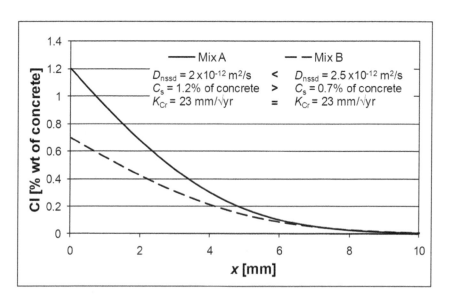

Figure 6.1 An example of coupled values of D_{nssd} and C_s.

presence of the strong external electric field and the short test duration, both of which tend to reduce the amount of bound, especially physically bound, chloride ions. Therefore, as pointed out by Tang (1996b), the parameter D_{nssm} describes the property of chloride transport under a condition of reduced chloride binding.

As D_{nssm} and D_{nssd} are obtained under completely different test conditions, their values may not be necessarily comparable. However, experimental results obtained from well-organised comparison tests, such the CHLOR-TEST project (Castellote and Andrade, 2005; Tang, 2005), and some other projects (Frederiksen *et al.*, 1997b; Tang and Sørensen, 2001), show that these two diffusion coefficients are, coincidentally, quite comparable (see Figure 6.2). However, some experimental data reported by other researchers (e.g. Salta *et al.*, 2006) do not lie within the standard deviation range shown in Figure 3.25, and a possible reason for this could be the different concrete ages in these tests. Taking into account the uncertainties in the measurements made using these test methods, it is reasonable to conclude that both test methods will provide similar transport parameter values if the effect of concrete age can be eliminated or reduced.

6.4.3 Results of the resistivity test

As discussed in Section 3.7.3, resistivity is, theoretically, inversely proportional to diffusivity. Practically, however, the all the ions in the pore

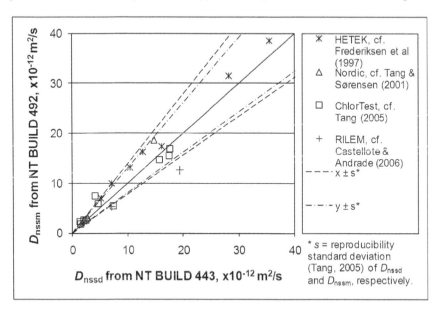

Figure 6.2 Relationship between the chloride transport parameters D_{nssm} and D_{nssd}. s is the reproducibility of the standard deviation of D_{nssm} and D_{nssd} (Tang, 2005).

solution, especially hydroxyl ions, contribute to the measured resistivity. A concrete containing a binder that is highly alkaline will, therefore, have a low resistivity, while a concrete containing pozzolanic additions, which have a low alkalinity, will often have a high resistivity. In order to convert the resistivity to the chloride diffusion coefficient, the chloride transference number (the ratio of the number of chloride ions to the total number of all ions) in the pore solution must be known, and it is not an easy task to estimate the total number of ions in the pore solution. Therefore, the results of the resistivity test can only be taken as an indirect measure of the chloride transport property. However, owing to its rapidity and simplicity, this test is a very efficient tool for undertaking quality control in the production of concrete, but an appropriate calibration is needed in order to establish empirical relationships between the resistivity and the chloride diffusion coefficient (see Figures 3.30 and 3.31).

6.5 Conclusions on prediction models

The following main conclusions can be drawn regarding the various prediction models (see Chapters 4 and 5).

Most chloride-ingress models are based on solutions to Fick's second law. Experimental data from laboratory or field exposure tests are curve-fitted against these solutions, and two or more parameters are derived by regressions analysis. The so-called 'chloride diffusivity' and the 'surface chloride content' are found to be time dependent, and recent models include these effects. This is, however, only a mathematical formulation. In practice, high-quality exposure data for long exposure periods are urgently needed. Without such data, the results obtained using chloride-ingress models will always contain uncertainty, and this is especially true for the extrapolation of age exponents to long exposure durations. A major improvement in the quality of results would be achieved if the time dependences of the diffusion coefficients and the surface chloride contents could be clearly explained, but to date no such explanations have been elucidated. An explanation of the time dependence will require models based on physics and chemistry, rather than on just empirical comparisons with measured exposure data.

Physical models, especially those that consider both all the ions present and convection, give theoretical predictions that most closely match the presently available measured data. Such models can be used to check the performance of more simple, empirical models. This process may explain some of the assumptions that are made in the simple models, and may improve the confidence in them. Physical models require a number of material parameters and a number of environmental parameters to be quantified, each as a function of several model parameters. In particular, the surface climatic conditions are extremely complicated to model. This requirement for a huge amount of input data, which are not readily available at present, and the limited user-friendliness of physical models, means that most of them will remain as research tools for the foreseeable future, and not be used in practical applications.

6.6 Acceptance criteria

6.6.1 How to set acceptance criteria

In normal cases, the presence of chloride ions does not in itself result directly in damage to concrete, but it does induce corrosion of any steel within the concrete. The service life of reinforced concrete structures exposed to chloride environments includes the periods of corrosion initiation and propagation. The former is related to chloride ingress, while the latter is related to the rate of corrosion. The period of corrosion initiation is a function of the chloride transport property, concrete cover, threshold chloride level, exposure environment and so on. Obviously, the chloride transport property is only one of several parameters that influence the initiation of corrosion. A relatively high diffusivity of a concrete can be compensated for by a thicker cover in order to obtain the desired resistance to corrosion initiation. Therefore, the acceptance criteria for values obtained from the proposed tests are dependent on many factors, and can be expressed by the following function:

$$D = f\left(x_{min}, t_L, C_{Cr}, K_{env}\right) \tag{6.1}$$

where D is the value from a proposed test, f is a function, x_{min} is the minimum thickness of the concrete cover, t_L is the desired period of corrosion initiation, C_{Cr} is the threshold chloride level, and K_{env} is the environmental factor, which includes the chloride load (e.g. surface concentration or content) and micro-climate (temperature, humidity and precipitation).

To set the acceptance criteria, the above function has to be solved; but to solve the function, proper models are needed. Different models for predicting the chloride ingress into concrete were evaluated in the CHLORTEST project (Nilsson, 2005), with collected in-field data at exposure times from 0.5 years up to 42 years (see Section 5.4). The results showed that Model 5 (the ClinConc model), which was previously calibrated using 10 years of traceable data from a field exposure site, provided a reasonably good benchmark, indicating the importance of calibrating a model against reliable long-term data. In the past decade, some models have been validated using field data taken from real concrete structures after 18–30 years in service within a chloride-containing environment (see Section 5.5). It can be concluded from the results of these validations, that the ClinConc model gives reasonably good predictions of the chloride ingress into concrete exposed in marine or de-icing salt road environments. Although validation against data obtained over 30 years for structures in service may still not be sufficient for the prediction of the service life of structures that are intended to last for hundreds of years, predictions made using a mechanistic model with a known extrapolation should still be much better than those made using empirical models. Therefore, if long-term field data of adequate quality are not available for the validation of simple empirical models, use of the ClinConc model is recommended instead. Nevertheless, it is always the user's responsibility to use D values obtained from the recommended tests and to choose the correct models when examining the service life design of a particular concrete structure.

6.6.2 Some examples

As a first example, consider the Coulomb values shown in Table 3.3, where the chloride ion penetrability is classified into five levels: negligible, very low, low, moderate and high. The pore solutions of concretes containing pozzolanic additions have a low hydroxyl ion concentration, due to secondary hydration, and therefore normally have lower Coulomb values, which will easily be classified as low, very low or even negligible chloride penetrability. Of course, many research results show that the diffusivity of concretes containing pozzolanic additions is lower than the diffusivity of concretes containing only Portland cement. However, the question arises, regarding the classification in Table 3.3, as to what portion of the reduction in the Coulomb value is attributable to the reduction in pH. On the other hand, the criteria in Table 3.3 may have little relationship to the service life of a concrete structure.

The Chinese standard JGJ/T 193–2009 classifies the resistance of concrete to chloride ingress as shown in Table 6.1. Again, the criteria for these classifications are not related to the service life of concrete structures.

From the point view of practical applications, it is important to correlate the laboratory-measured diffusivity with the minimum cover thickness. As the minimum cover thickness is dependent not only on the chloride diffusivity, but also on the chloride threshold level, the designed or expected service life, and the exposure environment, all these factors should be taken into account when establishing such a relationship. Tang (2004), through the study of

Table 6.1 Classification of concrete resistance to chloride ingress (according to JGJ/T 193–2009)

Resistance class	$D_{RCM}{}^1$ $(\times 10^{-12}\ m^2/s)$	Remarks
RCM-I	$D_{RCM} \geq 4.5$	Poor
RCM-II	$3.5 \leq D_{RCM} < 4.5$	Relatively poor
RCM-III	$2.5 \leq D_{RCM} < 3.5$	Relatively good
RCM-IV	$1.5 \leq D_{RCM} < 2.5$	Good
RCM-V	$D_{RCM} < 1.5$	Very good

Note
1 Measured at an age of 84 days using the RCM test (NT BUILD 492).

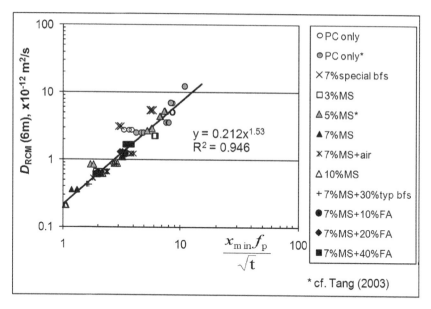

Figure 6.3 The relationship between the concrete diffusivity and the minimum cover thickness. BFS, blast furnace slag; FA, fly ash; MS, micro-silica; PC, Portland cement.

measured and modelled results, as shown in Figure 6.3, found the following relationship between the diffusivity and the minimum cover thickness:

$$D_{6m} = a\left(\frac{x_{min}f_p}{\sqrt{t_L}}\right)^b$$
(6.2)

where D_{6m} is the diffusivity $(\times 10^{-12}\,m^2/s)$ measured at age 6 months by the RCM test (NT BUILD 492), a and b are regression constants, x_{min} is the calculated minimum cover thickness (mm), t_L is the expected or designed service life (years), and f_p describes the effect of pozzolanic additives:

$$f_p = \begin{cases} 1 + \sum k_p M_p & \text{for } \sum k_p M_p \geq -0.5 \\ 0.5 & \text{for } \sum k_p M_p < -0.5 \end{cases}$$
(6.3)

where k_p is the activity factor of pozzolanic additives with regard to the chloride threshold and M_p is the mass ratio of the pozzolanic additive to cement. The values of k_p are as shown in Table 6.2.

In DIN EN 206 the exposure environments in the presence of chlorides are classified into six types. Owing to large variations in the environmental

Table 6.2 Values of k_p

Pozzolan	Silica fume	Fly ash	Typical blast furnace slag	Special blast furnace slag
k_p	−4.7	−1.4	−1.4	−4.7
Reference	Frederiksen (2000)		ConLife (Holt, 2004)	

parameters and the lack of valid field data, the relationship shown in Figure 6.3 was based mainly on modelling of the submerged seawater environment (XS2) and the field data from the Träslövsläge field site on the west coast of Sweden. To describe the effects of various exposure environments, an environmental factor $K_{x,\text{env}}$ can be added into Eq. 6.2:

$$x_{\min} = K_{x,\text{env}} \cdot \frac{\sqrt{t_L}}{f_p} \left(\frac{D_{6m}}{a} \right)^{\frac{1}{b}} \tag{6.4}$$

According the results shown in Figure 6.3, the constants are $a = 0.212$ and $b = 1.53$. Under the assumption of simple multiplication of various effects, $K_{x,\text{env}}$ can be expressed as:

$$K_{x,\text{env}} = k_{x,\text{ex}} \sqrt{\frac{c}{c_0}} \exp\left(\frac{E_D}{R} \left(\frac{1}{T_0} - \frac{1}{T} \right) \right) \tag{6.5}$$

where $k_{x,\text{ex}}$ is a factor depending on the exposure class; c and T are the chloride concentration and exposure temperature (K), respectively; c_0 and T_0 are the reference chloride concentration and temperature respectively (in this study the values at Träslövsläge, $c_0 = 14\,\text{g/l}$ and $T_0 = 284\,\text{K}$ (11°C)); E_D is the activation energy for chloride diffusivity; $E_b = 42\,000\,\text{J/mol}$; and R is the gas constant (8.314 J/(mol K)).

Based on the modelling results obtained in the EU project ConLife (Holt, 2004), the values of $k_{x,\text{ex}}$ given in Table 6.3 can be suggested. It should be noted that more in-field data are needed to verify these suggested values.

Thus by testing the chloride diffusivity at 6 months age using the RCM test, the minimum cover thickness can be calculated from Eq. 6.4 by taking into account the desired service life, the concrete mix and the exposure class.

Table 6.3 Suggested values of $k_{x,\text{ex}}$ for various exposure classes

DIN EN 206 exposure class	$k_{x,\text{ex}}$	Notes
XS1, XD1	0.4	Airborne chlorides
XD2	0.8	Mild chloride attack
XS2	1	Under seawater
XS3, XD3	1.3	Severe chloride attack

However, in practice, this calculated minimum cover thickness should be checked against the relevant national requirements. If the calculated value is less than the value specified in the national requirements, the latter value should be used.

Some calculated values of minimum cover thickness x_{min} for a service life of 100 years under various exposure conditions are given in Table 6.4. For the Swedish de-icing salt road environment, concrete containing ordinary structural Portland cement CEM I and a water/cement ratio of 0.4, it is possible to have a service life of 100 years if the cover thickness is greater than 45 mm and the chloride load corresponds to a concentration of 1.5 g/l. However, this type of concrete cannot withstand 100 years in the Swedish marine environment (XS2 or XS3), even with a cover thickness of 100 mm. Other types of concrete having lower diffusion coefficients are needed for this environment. For marine environments with a chloride concentration or temperature higher than that prevailing in Sweden, concretes with even lower diffusion coefficients are needed. This could be the reason why in the Chinese standard relatively low values of the chloride diffusion coefficient are used in the criteria (see Table 3.1), because both the chloride concentration and the water temperature in seawater to the south of China are higher than the corresponding values in Swedish seawater.

The above examples demonstrate again that both material properties and environmental characteristics must be taken into account when formulating acceptance criteria for the durability of concrete structures.

6.7 General remarks and recommendations for further progress

As stated in Chapter 1, the intention of this book is to summarise and critically appraise the current understanding of the fundamentals of chloride transport and the various models used to describe it, the tests that are commonly used and provide scientific explanations for the values obtained, and illustrate how models can be used to reliably predict the chloride transport in concrete in various exposure environments. As described, the science of chloride transport is extremely challenging, due to the varying nature of exposure regimens (and the associated environmental loads) and the cement matrixes through which the transport occurs. By summarising precision test data obtained from international test programmes, the repeatability and reproducibility of some of the most commonly used test methods were given, so that readers can use them to produce test data for prediction models. However, it must be remembered that a detailed description of the environment is essential to make best use of the models. The models themselves vary in nature, and some are better and scientifically more reliable than others. The discussion of the field data and associated modelling presented in this book clearly illustrate that further work is needed before any of the physical models are used routinely for the design of concrete structures in different exposure environments. However, by properly making use of both

Table 6.4 Calculated minimum cover thickness x_{min} for a service life of 100 years under various exposure conditions

Environment	De-icing salt road			Marine		
DIN EN 206 exposure class	XD1	XD2	XD3	XS1	XS2	XS3
$k_{x,ex}$	0.4	0.8	1.3	0.4	1	1.3
c		1.5			14	14 28
T (°C)		10			11	11 18
$K_{x,env}$	0.123	0.246	0.4	0.4	1	1.84 2.82
Binder	Portland cement CEM I					CEM I + 5% silica fume
Water/binder ratio		0.4			0.4	0.35 < 0.35 or by other means to reduce D_{6m}
D_{6m} (×10⁻¹² m²/s)		8.6			4.4	3 1.5 1.5 1
x_{min} (mm)	14	28	45	45	112	97 126 98 62 88 104

the environmental data, the material composition and the test parameters, it is possible to specify concretes and the depth of cover required to prevent chloride-induced corrosion under the different exposure conditions specified in DIN EN 206.

The authors believe that this book will encourage both researchers and practising engineers to deal with the chloride-transport phenomenon differently, due to improved knowledge based on the information presented in various chapters. However, readers are advised to consult experts on this important topic before further research is undertaken, so that any further data collected can contribute to filling some of the existing gaps in the knowledge of this subject.

Appendix
Test methods for determining the resistance of concrete to chloride ingress

A1 Introduction

Chloride ions can induce corrosion of the reinforcement contained in concrete. Reinforced concrete structures exposed to environments containing chlorides need to be durable and have an adequate resistance to the ingress of chlorides. Owing to the important role of chloride ingress with regard to the durability of concrete structures, many different test methods have been developed. Some of the commonly used test methods were evaluated in the EU project CHLORTEST under the 5th Frame Programme (GRD1-2002-71808/G6RD-CT-2002-00855). Based on the results of a round-robin test programme in this project, the CHLORTEST consortium recommended three methods for testing the resistance of concrete to chloride ingress: the immersion test, the rapid chloride migration (RCM) test and the resistivity test.

It is desirable, especially for new constituents or new concrete compositions, to test the resistance of a concrete to chloride ingress before it is used in construction. This applies also to concrete mixes, concrete products, precast concrete products, concrete members and concrete *in situ*. The tests can be used either to supply quantitative data for use in the design of durability, or to assess the quality of concrete during its production or use in construction.

A2 Terms and definitions

For the purposes of these test methods, the following terms and definitions apply:

as-cast surface The surface of a concrete structure exposed to the chloride-containing environment.

chloride penetration The ingress of chloride ions into concrete due to exposure to external chloride sources.

diffusion The movement of molecules or ions under a concentration gradient or, more strictly, chemical potential, from an area of high concentration to an area of low concentration.

maturity-day A concrete of 28 maturity-days has developed a maturity corresponding to curing in 28 days at 20°C.

migration The movement of ions under the action of an external electric field.

profile grinding Grinding off concrete powder in thin successive layers from a test concrete using a dry process.

resistivity The electrical resistance per unit length and per unit reciprocal cross-sectional area of concrete at a specified temperature.

surface-dry condition A surface condition achieved by drying the water-saturated test specimen with a clean cloth or similar, leaving the test specimen damp but not wet.

A3 Test specimens in general

Drilled cores or cast cylinders can be used as test specimens. The specimen must be representative of the concrete and/or structure in question. The test specimen diameter is usually 100 mm, or at least four times the maximum aggregate size. At least three specimens should be tested. If cast cylinders or cores drilled from a cast cube are used as specimens, the casting and curing procedures should be in accordance with relevant standards (e.g. DIN EN 12390-2).

In normal cases, the concrete should be hardened to at least 28 maturity-days for testing. As the concrete age has a significant effect on the transport of chloride ions, the date of manufacture of the concrete and the date of testing should always be noted in the report. If the concrete temperature during hardening was outside the range 10–30°C, this must also be noted in the report.

A4 Immersion test

A4.1 Principle

One plane surface of a water-saturated concrete specimen is exposed to water containing sodium chloride. After a specified exposure time, thin layers are ground off parallel to the exposed face of the specimen, and the chloride content profile is measured. The apparent chloride diffusion coefficient and the chloride content at the exposed surface are calculated by curve-fitting the measured profile to the error function solution to Fick's second law. A penetration parameter combining the influence of the diffusion coefficient, the surface chloride content, the initial chloride content and a reference chloride content is calculated in order to facilitate comparison of the test results obtained for different types of concrete.

A4.2 Reagents and equipment

A4.2.1 Reagents

- distilled or demineralised water;
- calcium hydroxide ($Ca(OH)_2$), technical quality;
- sodium chloride (NaCl), chemical quality;
- two-component, solvent free (chloride-ion and water-vapour diffusion proof) polyurethane or epoxy-based paint (membrane);
- chemicals for chloride analysis as required by the test method employed (see Section A4.4.5).

A4.2.2 Equipment

- equipment for cutting specimens, e.g. a water-cooled diamond saw;
- balance, with an accuracy better than ±0.01 g;
- thermometer or thermocouple, with a readout device capable of reading to ±1°C;
- plastic container with a tight-fitting lid;
- slide calliper with a precision of ±0.1 mm;
- ruler with a minimum scale of 1 mm;
- equipment for grinding off and collecting concrete powder from thin (<2 mm) concrete layers;
- standard sieve, mesh width 1.0 mm;
- equipment for chloride analysis as required by the test method employed (see Section A4.4.5).

A4.3 Preparation of the test specimen

A4.3.1 Test specimen

If drilled cores are used, cut an approximately 70 mm thick slice from the outermost portion of each core as the test specimen. The surface 10 mm below the as-cast surface is the surface to be exposed to the chloride solution (see Section A4.3.2, point 4).

If cast cylinders are used, cut an approximately 70 mm thick slice from the central part of each cylinder as the test specimen. The end surface that is nearest to the as-cast surface is the surface to be exposed (see Section A4.3.2, point 4).

A slice at least 20 mm thick is cut from the remainder of the drilled core or cast cylinder for the measurement of the initial chloride content (see Section A4.4.5, point 2).

NOTE 1: The term 'cut' here means to saw perpendicularly to the axis of a core or cylinder, using a water-cooled diamond saw at a slow speed in order to eliminate, or at least reduce, micro-cracking during cutting.

NOTE 2: It is very important that the test is made on the concrete between the surface and the layer of reinforcement, because it is here the protection against chloride penetration is needed. Furthermore, the quality of the concrete in this particular area can deviate from that of the rest of the concrete. The outermost approximately 10 mm of concrete should be removed (see Section A4.3.2, point 4) to ensure that the measurement is made in an area having a relatively constant cement matrix content and unaffected by surface treatments, lack of curing, etc., *unless the measurement is for the skin effect.*

A4.3.2 Pre-conditioning and coating

1   The test specimen is immersed in a saturated $Ca(OH)_2$ solution at 20–25°C in a tightly closed plastic container. The container must be filled to the top in order to minimise carbonation of the liquid. The next day the mass, in a surface-dry condition, is determined by weighing the test specimen.
2   Storage in the saturated $Ca(OH)_2$ solution is continued until the mass of the specimen, in the surface-dry condition, does not change by 0.1% by mass when measured at an interval of at least 24 hours.
3   All surfaces of the test specimen at room temperature, except the one to be used for exposure, are dried to a stable white-dry condition, and a thick epoxy or polyurethane coating is applied on all white-dry surfaces. It must be ensured that the method of application and hardening prescribed by the supplier of the coating material is observed.
4   When the coating has hardened, the outermost approximately 10 mm thick layer from the surface to be used for exposure is cut off, unless the measurement is for the skin effect, and then the remainder is immersed in the $Ca(OH)_2$ solution overnight.

A4.4 Test procedures

A4.4.1 Exposure liquid

An aqueous NaCl solution is prepared with a concentration of 165 ± 1 g NaCl per litre of solution. This exposure liquid is used for 5 weeks, and then replaced with fresh NaCl solution. The NaCl concentration of the solution is checked at least before and after use.

A4.4.2 Exposure temperature

The temperature of the water bath is kept in the range 20–25°C, with a target average temperature of 23°C during the immersion. The temperature is measured at least once a day.

A4.4.3 Exposure

1   The surface-dry specimens are placed in the container in such a way that the exposure surface is vertically exposed.
2   The container is filled with the exposure liquid. It is important that the container is completely filled with liquid and closed tightly. The ratio of the exposed area (cm²) to the volume of exposure liquid (litres) is maintained in the range 20–80.
3   The container is stored in the room at the temperature specified in Section A4.4.2.
4   The exposure is continued for at least 35 days, but in normal cases does not exceed 40 days. The date and time of the start and end of the exposure is recorded to ±10 minutes.

A4.4.4 Profile grinding

1   After the exposure, the specimen is lightly rinsed with tap water.
2   Any excess water is wiped off the surfaces of the specimen.
3   Material is ground off in layers parallel to the exposed surface, ensuring that the grinding area is within a diameter approximately 10 mm less than the full diameter of the core from the exposure surface. Layers are ground off at small depth intervals, depending on what is appropriate for the grinding tool used. However, the depth between each layer must always be more than three times the maximum size of the aggregates. This removes the risk of edge effects and coating contamination of the samples.
4   At least eight layers are ground off. The thickness of the layers is adjusted according to Table A1 or the expected chloride profile, in order to ensure a minimum of six points in the profile having a chloride content higher than the initial value (see Section A4.4.5, point 2). However, the outermost layer must always have a minimum thickness 1.0 mm.

Table A1 Recommended depth intervals (mm) for powder grinding

Depth No.	Water/binder ratio						
	0.25	*0.30*	*0.35*	*0.40*	*0.50*	*0.60*	*0.70*
1	0–1	0–1	0–1	0–1	0–1	0–1	0–1
2	1–2	1–2	1–2	1–3	1–3	1–3	1–5
3	2–3	2–3	2–3	3–5	3–5	3–6	5–10
4	3–4	3–4	3–5	5–7	5–8	6–10	10–15
5	4–5	4–6	5–7	7–10	8–12	10–15	15–20
6	5–6	6–8	7–9	10–13	12–16	15–20	20–25
7	6–8	8–10	9–12	13–16	16–20	20–25	25–30
8	8–10	10–12	12–16	16–20	20–25	25–30	30–35

Notes
For concrete with pozzolanic additions such as fly ash, slag and silica fume, the depth intervals applied for any given water/binder ratio should be those corresponding to the preceding column (i.e. one column to the left of the given water/binder ratio); e.g. for slag cement concrete with w/b = 0.4, the depth intervals in the column for w/b = 0.35 should be applied.

5   A sample of at least 5 g of dry concrete dust is obtained from each layer. For each sample of concrete dust collected, the depth below the exposed surface is calculated as the average of four uniformly distributed measurements made using a slide calliper.

A4.4.5 Measurement of chloride content

1   The acid-soluble chloride content of the samples is determined to three decimal places in accordance with the relevant standard methods (e.g. EN 14629 or a similar method having the same or better accuracy).
2   From the concrete slice (see Section A4.3.1, point 3), a representative sample of approximately 20 g is prepared by grinding or crushing until the material passes through a 1 mm standard sieve. The initial chloride content is determined using the same method as in Section A4.4.5, point 1.

A4.5 Expression of results

The values of the apparent non-steady-state chloride diffusion coefficient D_{nssd} and the apparent surface chloride content C_s are determined by curve-fitting the measured chloride profile to Eq. A1 according to the principle of least squares, as illustrated in Figure A1:

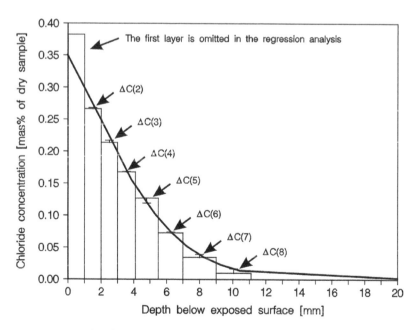

Figure A1 Example of regression analysis for curve-fitting (NT BUILD 443, 1994).

$$C(x, t) = C_s - (C_s - C_i)\, \text{erf}\left(\frac{x}{\sqrt{4D_{nssd}t}}\right) \tag{A1}$$

where:

D_{nssd} is the apparent non-steady-state diffusion coefficient (m²/s)
x is the average depth at which the chloride sample was ground (m)
t is the duration of immersion (s)
$C(x, t)$ is the chloride content at depth x (mass % of sample)
C_s is the apparent surface chloride content (mass % of sample)
C_i is the initial chloride content (mass % of sample;
erf is the error function, as defined by Eq. A2.

$$\text{erf}(z) = \frac{2}{\sqrt{\pi}} \int_0^z e^{-u^2}\, du \tag{A2}$$

The penetration parameter K_{Cr} is calculated using Eq. A3, with the above obtained values of C_i, C_s and D_{nssd}, and assuming that $C_r = 0.05\%$ by mass of sample, unless another value is required.

$$K_{Cr} = 2\sqrt{D_{nssd}}\ \text{erf}^{-1}\left(\frac{C_s - C_r}{C_s - C_i}\right) \tag{A3}$$

where: erf^{-1} is the inverse of the error function.

NOTE: K_{Cr} is defined only when $C_s > C_r > C_i$.

A4.6 Test report

The test report should contain at least the following information:

- reference to this standard method;
- the origin, size and marking of the specimens;
- the concrete identification;
- the date of manufacture of the concrete;
- details of any deviation from the test method;
- the test results, including the specimen dimensions, concrete age at exposure, exposure temperatures, chloride concentration of the exposure liquid, exposure duration, data of chloride content versus depth, the curve-fitted diffusion coefficient and surface content, and the calculated penetration parameter;
- details of any observed or suspected defects (cracks, large pore holes, foreign bodies, etc.) in the specimen.

A5 Rapid chloride migration (RCM) test

A5.1 Principle

An external electrical field is applied axially across the specimen, which forces the chloride ions outside to migrate into the specimen. After a certain test duration, the specimen is axially split or dry-cut, and a silver nitrate solution is sprayed onto one of the freshly split or cut sections. The chloride penetration depth is then measured from the visible grey-silver chloride precipitation, after which the chloride migration coefficient is calculated from this penetration depth.

A5.2 Reagents and equipment

A5.2.1 Reagents

- distilled or demineralised water;
- sodium chloride (NaCl), chemical quality;
- sodium hydroxide: (NaOH), chemical quality;
- silver nitrate (AgNO$_3$), chemical quality;
- chemicals for chloride analysis, as required by the test method employed (optional, see Section A5.4.6).

A5.2.2 Equipment

- equipment for cutting specimens, e.g. a water-cooled saw;
- vacuum container, capable of containing at least three specimens;
- vacuum pump, capable of maintaining a pressure of < 50 mbar (5 kPa) in the container, e.g. a water-jet pump;
- migration set-up: one design is shown in Figure A2, which includes the following parts:

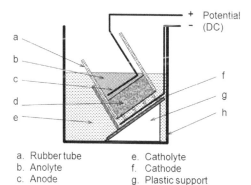

a. Rubber tube
b. Anolyte
c. Anode
d. Specimen
e. Catholyte
f. Cathode
g. Plastic support
h. Plastic box

Figure A2 One arrangement of the RCM test set-up.

- silicone rubber tube, with an inner diameter suitable for the diameter of the specimens, of thickness 5–7 mm and length approximately 150 mm;
- stainless steel clip, 20 mm wide and with a diameter range suitable for the rubber tube (Figure A3);

Figure A3 Stainless-steel clip.

- catholyte reservoir, a plastic box, $370 \times 270 \times 280$ mm (length × width × height);
- plastic support (Figure A4);

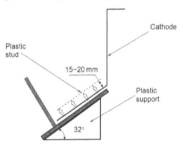

Figure A4 Plastic support and cathode.

- cathode, a stainless steel plate (Figure A4), about 0.5 mm thick;
- anode, a stainless steel mesh or plate with holes (Figure A5), about 0.5 mm thick.

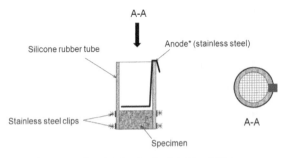

* The diameter of anode is 5 mm less than that of the specimen.

Figure A5 Rubber tube assembled with the specimen, clips and anode. The diameter of the anode is 5 mm less than that of the specimen.

NOTE: Other designs are acceptable, provided that the temperature of the specimen and solutions can be maintained in the range of 20–25°C during the test (see Section A5.4.2).

- power supply, capable of supplying a 0–60 V DC regulated voltage with an accuracy of ±0.1 V;
- ammeter, capable of displaying current to ±1 mA;
- thermometer or thermocouple, with a readout device capable of reading to ±1°C;
- any suitable device for splitting (e.g. compression machine) or dry-cutting (e.g. air-cooled diamond saw) the specimen;
- spray bottle;
- slide calliper with a precision of ±0.1 mm;
- ruler with a minimum scale of 1 mm;
- equipment for chloride analysis as required by the test method employed (optional, see Section A5.4.6).

A5.3 Preparation of the test specimen

A5.3.1 Test specimen

- If drilled cores are used, the outermost approximately 10 mm thick layer should be cut off, unless the measurement is for the skin effect, and the next 50 ± 2 mm thick slice should be cut from each core as the test specimen. The end surface that was nearest the outermost layer is the one to be exposed to the chloride solution (catholyte).
- If cast cylinders are used, a 50 ± 2 mm thick slice should be cut from the central portion of each cylinder as the test specimen. The end surface that was nearest to the as-cast surface is the one to be exposed to the chloride solution (catholyte).
- The thickness is measured using a slide calliper and read to 0.1 mm.

NOTE: The term 'cut' here means to saw perpendicularly to the axis of a core or cylinder, using a water-cooled diamond saw operated at a slow speed.

A5.3.2 Pre-conditioning

After cutting, any burrs from the surfaces of the specimen should be brushed and washed away, and the excess water wiped off the surfaces of the specimen. When the specimens are surface dry, they are placed in the vacuum container for vacuum treatment. Both end surfaces must be exposed. The absolute pressure in the vacuum container is reduced to within the range 10–50 mbar (1–5 kPa) within a few minutes. The vacuum is maintained for 3 hours and then, with the vacuum pump still running, the container is filled with the distilled or demineralised water so as to immerse all the specimens. The vacuum is maintained for a further hour before air is allowed to re-enter the container. The specimens are kept in the solution for 18 ± 2 hours.

A5.4 Test procedures

A5.4.1 Catholyte and anolyte

The catholyte solution is 10% NaCl by mass in tap water (100 g NaCl in 900 g water, about 2 N), and the anolyte solution is 0.3 M NaOH in distilled or demineralised water (approximately 12 g NaOH in 1 litre water). The solutions are stored at a temperature of 20–25°C.

NOTE: It is important to use distilled or demineralised water for the anolyte solution to prevent corrosion damage to the anode.

A5.4.2 Temperature

The temperatures of the specimen and solutions are maintained in the range 20–25°C during the test.

A5.4.3 Preparation of the test

1   The catholyte reservoir is filled with about 12 litres of 10% NaCl solution.
2   The rubber tube is fitted on the specimen as shown in Figure A5 and secured with two clips. If the curved surface of the specimen is not smooth, or there are defects on the curved surface that could result in significant leakage, a line of silicone sealant can be applied to improve the tightness.
3   The specimen is placed on the plastic support in the catholyte reservoir (see Figure A2).
4   The tube above the specimen is filled with about 300 ml anolyte solution (0.3 M NaOH) in order to cover the specimen surface with a liquid layer of at least 3 mm.
5   The anode is immersed in the anolyte solution.
6   The cathode is connected to the negative pole and the anode to the positive pole of the power supply.

A5.4.4 Migration test

1   After the power is turned on, the voltage is preset at 30 V, and the initial current through each specimen is recorded.
2   The voltage is adjusted if necessary (as shown in Table A2). After adjustment, the value of the initial current is recorded again.

Table A2 Test voltage and duration for concrete specimens having a normal binder content

Initial current, $I_{30 V}$ (with 30 V) (mA)	Applied voltage, U (after adjustment) (V)	Possible new initial current, I_0 (mA)	Test duration, t (hours)
$I_0 < 5$	60	$I_0 < 10$	96 or longer
$5 \leq I_0 < 10$	60	$10 \leq I_0 < 20$	48
$10 \leq I_0 < 15$	60	$20 \leq I_0 < 30$	24
$15 \leq I_0 < 20$	50	$25 \leq I_0 < 35$	24
$20 \leq I_0 < 30$	40	$25 \leq I_0 < 40$	24
$30 \leq I_0 < 40$	35	$35 \leq I_0 < 50$	24
$40 \leq I_0 < 60$	30	$40 \leq I_0 < 60$	24
$60 \leq I_0 < 90$	25	$50 \leq I_0 < 75$	24
$90 \leq I_0 < 120$	20	$60 \leq I_0 < 80$	24
$120 \leq I_0 < 240$	15	$60 \leq I_0 < 120$	24
$240 \leq I_0 < 600$	10	$80 \leq I_0 < 200$	24
$I_0 > 600$	10	$I_0 > 200$	6

Notes
1 The values in the table are based on specimens of 100 mm diameter. For specimens of another diameter d (mm), correct the measured current by multiplying by a factor of $(100/d)^2$ in order to be able to use the values in the table.
2 For specimens having a special binder content, such as repair mortars or grouts, correct the measured current by multiplying by a factor approximately equal to the ratio of normal binder content to the actual binder content in order to be able to use the values in table.

3   The initial temperature in each anolyte solution is measured using a thermometer or thermocouple.
4   The appropriate test duration is chosen according to the initial current (see Table A2).
5   The final current is recorded and the temperature is measured before the test is terminated.

A5.4.5 Measurement of the chloride penetration depth

1   The specimen is removed by following the reverse of the procedure in Section A5.4.3. A wooden rod is often helpful for removing the rubber tube from the specimen.
2   The specimen is rinsed with tap water.
3   The excess water is wiped off the surfaces of the specimen.
4   The specimen is axially split or dry-cut into two pieces. The piece having the split or cut section more nearly perpendicular to the end surfaces is chosen for the measurement of the penetration depth, and the other piece is kept for the chloride content analysis (optional). If there is no requirement for chloride content analysis, both pieces can be used for the penetration depth measurement.
5   The 0.1 M silver nitrate solution is sprayed onto the freshly split or cut section.

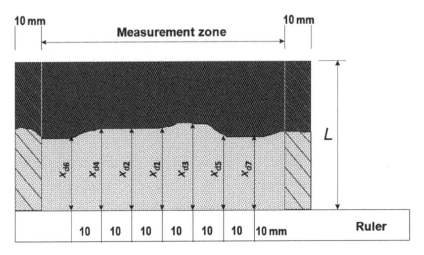

Figure A6 Illustration of the measurement of chloride penetration depths.

6   When the grey-silver chloride precipitation on the split or cut surface is clearly visible (after about 15 minutes), the penetration depth is measured using the slide calliper and a suitable ruler, from the centre to both edges at intervals of 10 mm (see Figure A6) to obtain seven depths. The depth is read to 0.1 mm.

NOTE 1: The term 'dry-cut' here means to saw the specimen using an air-cooled diamond saw.

NOTE 2: If the penetration front to be measured is obviously blocked by the aggregate, make the measurement at the nearest front where there is no significant blocking by aggregate; alternatively, ignore this depth if there are more than five valid depths.

NOTE 3: If there is a significant defect in the specimen that results in a penetration front much larger than the average, ignore this front, but note or photograph it and report the condition.

NOTE 4: To avoid the edge effect due to a non-homogeneous saturation or possible leakage, do not make any depth measurements in the zone within about 10 mm from the edge (see Figure A6).

A5.4.6 Measurement of the surface chloride content (optional)

1   From the unused axially split or cut specimen, cut an approximately 5 mm thick slice parallel to the end surface that was exposed to the chloride solution (catholyte).

2   The chloride content in the slice is determined in accordance with relevant standard methods (e.g. EN 14629, or a similar method with the same or better accuracy).

NOTE 1: Information about the chloride-binding capacity of the tested material may be estimated from the surface chloride content.

NOTE 2: The thickness of the slice should always be less than the minimum penetration depth.

A5.5 *Expression of results*

The non-steady-state migration coefficient is calculated as:

$$D_{nssm} = \frac{RT}{zFE} \cdot \frac{x_d - \alpha\sqrt{x_d}}{t} \tag{A4}$$

where

$$E = \frac{U-2}{L} \tag{A5}$$

and

$$\alpha = 2\sqrt{\frac{RT}{zFE}}\,\text{erf}^{-1}\left(1 - \frac{2c_d}{c_0}\right) \tag{A6}$$

D_{nssm} is the non-steady state migration coefficient(m^2/s)
z is the absolute value of the ion valence (for chloride, $z = 1$)
F is the Faraday constant ($= 9.648 \times 10^4$ J/(V mol))
U is the absolute value of the applied voltage (V)
R is the gas constant ($= 8.314$ J/(K mol))
T is the average value of the initial and final temperatures in the anolyte solution (K)
L is the thickness of the specimen (m)
x_d is the penetration depth (m)
t is the test duration (s)
erf^{-1} is the inverse of the error function
c_d is the chloride concentration at which the colour changes (≈ 0.07 N for ordinary Portland cement concrete)
c_0 is the chloride concentration in the catholyte solution (≈ 2 N).
    As

$$\xi = \text{erf}^{-1}\left(1 - \frac{0.07}{c_0}\right) = 1.28$$

the following simplified equation can be used:

$$D_{\text{nssm}} = \frac{0.0239(273+T)L}{(U-2)t}\left(x_d - 0.0238\sqrt{\frac{(273+T)Lx_d}{U-2}}\right) \qquad \text{(Eq. A7)}$$

where:

D_{nssm} is the non-steady-state migration coefficient ($\times 10^{-12}\,\text{m}^2/\text{s}$)
U is the absolute value of the applied voltage (V)
T is the average value of the initial and final temperatures in the anolyte solution (°C)
L is the thickness of the specimen (mm)
x_d is the penetration depth (mm)
t is the test duration (h).

A5.6 *Test report*

The test report should contain at least the following information:

- reference to this standard method;
- the origin, size and marking of the specimens;
- the concrete identification;
- the date of manufacture of the concrete;
- any deviation from the test method;
- the test results, including the specimen dimensions, concrete age at testing, applied voltage, initial and final currents, initial and final temperatures, test duration, average and maximum penetration depth, average and maximum migration coefficient, and individual data of penetration depths;
- any observed or suspected abnormal penetration front due to defects (cracks, large pore holes, foreign bodies etc.) in the specimen;
- optional information about the surface chloride content.

A6 Resistivity test

A6.1 *Principle*

The test is based on the measurement of the electrical resistance of a concrete sample by means of current lines parallel to the base of the sample.

A6.2 Equipment

- equipment for cutting specimens, e.g. a water-cooled diamond saw;
- vacuum container, capable of containing at least three specimens;
- vacuum pump, capable of maintaining a pressure of < 50 mbar (5 kPa) in the container, e.g. a water-jet pump;
- LCR meter, with a precision of 0.5%, able to measure resistance at a frequency of 1 kHz.

NOTE: It is acceptable to use an AC source with two external multimeters to measure the current and voltage drop.

- electrodes: metallic nets (openings < 2 mm) or plates, made of steel, copper or any other good conductive metal, and free of superficial impurities (deposits, rust, oxides, etc.), with a diameter equal to that of the specimen;
- contact sponges: two pieces of thin sponge (thickness < 5 mm) with a diameter equal to that of the electrodes;
- weight object: made of non-conducting material, with a mass of about 2 kg;
- ruler with a minimum scale of 1 mm.

A6.3 Preparation of the test specimen

A6.3.1 Test specimen

1   If drilled cores are used, cut off the outermost approximately 10 mm thick layer, and cut the next 50 ± 2 mm thick slice from each core as the test specimen.
2   If cast cylinders are used, cut a 50 ± 2 mm thick slice from the central portion of each cylinder as the test specimen.
3   The thickness is measured using a slide calliper and read to 0.1 mm.

NOTE: The term 'cut' here means to saw perpendicularly to the axis of a core or cylinder, using a water-cooled diamond saw at a slow speed.

A6.3.2 Pre-conditioning

A pre-conditioning procedure similar to that described in Section A5.3.2 can be used.

NOTE: If the specimens are cured for the entire time in the water, and no apparent drying has occurred during transport, drilling, cutting or other preparation procedures, the vacuum procedure can be skipped.

A6.4 Test procedures

A6.4.1 Testing room

The testing room shall have a temperature of $20\pm2°C$ and a relative humidity $RH\geq45\%$.

A6.4.2 Measurement arrangement

The measurement arrangement is as shown in Figure A7.

A6.4.3 Measurement of the ohmic resistance

1   The sponges are wetted with tap water and allowed to drip slightly to remove the excess water.
2   The excess water is wiped away from the surfaces of the samples.
3   The sponges, specimen, electrodes and the 2 kg weight are positioned as shown in Figure A7.
4   The resistance between the two electrodes is measured using the LCR meter at a frequency of 1 kHz, and noted as R_{s+sp}.
5   The specimen is removed, but the sponges, electrodes and 2 kg weight are left in place.
6   The resistance between the two electrodes is again measured using the LCR meter at a frequency of 1 kHz, and noted as R_{sp}.

A6.5 Expression of results

The resistivity is calculated as:

$$\rho = \frac{\pi d^2}{4000L}\left(R_{s+sp} - R_{sp}\right) \tag{A8}$$

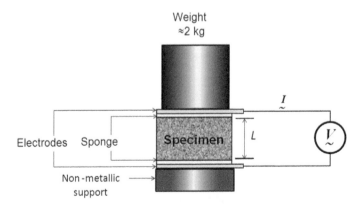

Figure A7 Arrangement for measuring the resistivity using the direct method.

where:

ρ is the resistivity of the concrete (Ωm)
d is the diameter of the specimen (mm)
L is the thickness of the specimen (mm)
R_{s+sp} is the resistance measured with the specimen and the sponges (Ω)
R_{sp} is the resistance measured with the sponges only (Ω).

A6.7 *Test report*

The test report should contain at least the following information:

- reference to this standard method;
- the origin, size and marking of the specimens;
- the concrete identification;
- the date of manufacture of the concrete;
- any deviation from the test method;
- the test results, including the specimen dimensions, concrete age at testing, temperature, imposed current, drops in potential and resistivity;
- any observed or suspected defects (cracks, large pore holes, foreign bodies etc.) in the specimen.

A7 Precision data

Based on the inter-laboratory test done in the EU project ChlorTest (GRD1-2002-71808/G6RD-CT-2002-00855) with three replicate specimens and 12 laboratories, the values for the repeatability and reproducibility of the three test methods described in Sections A4, A5 and A6 are listed in Table A3. These values may be used to assess the accuracy of data from any laboratory for these three test methods.

Table A3 Precision data for various test methods

Test method	Parameter	Repeatability, s_r	Reproducibility, s_R
Immersion test	D_{nssd}	$s_r = 0.201D_{nssd}$ ($R^2 = 0.94$)	$s_R = 0.283D_{nssd}$ ($R^2 = 0.99$)
	C_s	$s_r = 0.177C_s$ ($R^2 = 0.14$)	$s_R = 0.223C_s$ ($R^2 = 0.41$)
	K_{Cr}	$s_r = 0.090K_{Cr}$ ($R^2 = 0.76$)	$s_R = 0.135K_{Cr}$ ($R^2 = 0.88$)
Rapid chloride migration (RCM) test	D_{nssm}	$s_r = 0.152D_{nssm}$ ($R^2 = 0.86$)	$s_R = 0.236D_{nssm}$ ($R^2 = 0.93$)
Resistivity test	ρ	$s_r = 0.105\rho$ ($R^2 = 0.97$)	$s_R = 0.251\rho$ ($R^2 = 998$)

A8 Modifications of the test procedures

The major modifications of the test procedures described in Sections A4, A5 and A6, as compared with the previous versions, are summarised in Table A4. This summary is intended to highlight to readers who are familiar with previous versions the significant modifications made to the procedures before carrying out the precision test in the EU project.

Table A4 Major modifications in the proposed test methods

Proposed method	Previous version
Immersion test	*NT BUILD 443*
• Cut off approximately the outermost 10 mm layer *after* limewater saturation and coating, in order to prevent the exposure surface undergoing calcium densification	• Cut off approximately the outermost 10 mm layer *before* limewater saturation and coating
• Exposure duration '*at least* 35 days, *but in normal cases not exceed 40 days*', in order to reduce the age effect in normal tests	• Exposure duration '*at least* 35 days'
Rapid chloride migration (RCM) test	*NT BUILD 492*
• Vacuum saturation with *distilled or demineralised water* to simplify the procedure, because during this short saturation period there should be no significant leaching problem	• Vacuum saturation with *saturated limewater*
• After the migration test, '*split or dry-cut* the specimen', in order to assist the laboratory that may have no machine for splitting and to make the depth measurement easier	• After the migration test, '*split* the specimen'
Resistivity test	*Version used in the EU project CHLORTEST*
• The thickness of the specimen should be 50±2 mm, in order to reduce the size effect	• No size specification
• Vacuum to a pressure of 'less than *50 mbar (5 kPa) in the container*', for ease of testing, especially when specimens are moist or the container has been filled with water	• Vacuum to a pressure of 'less than 1 mmHg'[1]

Notes
1 Corresponding to 0.76 mbar or 0.076 kPa. This pressure was specified in the earlier version of AASHTO T277, but can hardly be achieved without a special vacuum pump, especially when specimens are moist or the container has been filled with water.

Bibliography

AASHTO T259 (2002), *Standard Method of Test for Resistance of Concrete to Chloride Ion Penetration*. American Association of State Highway and Transportation Officials, Washington, DC.

AASHTO TP 64-03 (2003), *Predicting Chloride Penetration of Hydraulic Cement Concrete by the Rapid Migration Procedure*. American Association of State Highway and Transportation Officials, Washington, DC.

AASHTO T277 (2007), *Standard Method of Test for Electrical Indication of Concrete's Ability to Resist Chloride Ion Penetration*. American Association of State Highway and Transportation Officials, Washington, DC.

Akita, H. and Fujiwara, T. (1995), Water and salt movement within mortar partially submerged in salty water. In: *Proceedings of the International Conference on Concrete under Severe Conditions*, August 1995, Sapporo, pp. 645–654. E&FN Spon, London.

Alisa, M., Andrade, C., Gehlen, C., Rodriguez, J. and Vogels, R. (1998), *Modelling of Degradation, DuraCrete: Probabilistic Performance based Durability Design of Concrete Structures*. EU Project (Brite EuRam III), BE95-1347, Report R 4/5.

Andrade, C. (1993), Calculation of chloride diffusion coefficients in concrete from ionic migration measurements. *Cement and Concrete Research*, 23(3):724–742.

Andrade, C. (2005), *Task 4.3: Benchmarking of Models (Internal Working Report), ChlorTest: Resistance of Concrete to Chloride Ingress – From Laboratory Tests to In-field Performance*. EU Project (5th FP GROWTH), G6RD-CT-2002-00855.

Andrade, C. and Page, C.L. (1986), Pore solution chemistry and corrosion in hydrated cement systems containing chloride salts: a study of cation specific effects. *British Corrosion Journal*, 21(1):49–53.

Andrade C., Alonso C., Arteaga A. and Tanner P. (2000a), Methodology based on the electrical resistivity for the calculation of reinforcement service life. In: *Proceedings of the 5th CANMET/ACI International Conference on Durability of Concrete*, 4–9 June 2000, Barcelona (ed. V.M. Malhotra). Supplementary paper, pp. 899–915.

Andrade, C., Castellote, M., Alonso, C. and Gonzalez, C. (2000b), Non-steady-state chloride diffusion coefficients obtained from migration and natural diffusion tests. Part I: Comparison between several methods of calculation. *Materials and Structures*, 33(225):21–28.

Arya, C. and Newman, J.B. (1990), An assessment of four methods of determining the free chloride content of concrete. *Materials and Structures*, 23(137):319–330.

Arya, C., Buenfeld, N.R. and Newman, J.B. (1987), Assessment of simple methods of determining the free chloride ion content of cement paste. *Cement and Concrete Research*, 17(6):907–918.

ASTM C1543 (2010), *Standard Test Method for Determining the Penetration of Chloride Ion into Concrete by Ponding*. ASTM International, West Conshohocken, PA.

ASTM C1556 (2004), *Standard Test Method for Determining the Apparent Chloride Diffusion Coefficient of Cementitious Mixtures by Bulk Diffusion*. ASTM International, West Conshohocken, PA.

ASTM C1202 (2010), *Standard Test Method for Electrical Indication of Concrete's Ability to Resist Chloride Ion Penetration*. ASTM International, West Conshohocken, PA.

Basheer, P.A.M., Andrews, R.J., Robinson, D.J. and Long, A.E. (2005), 'PERMIT' ion migration test for measuring the chloride ion transport of concrete on site. *NDT&E International*, 38(3):219–229.

Bentz, E. and Thomas, M. (1999), *User manual of Life 365: Computer program for predicting the service life and life-cycle costs of reinforced concrete structures exposed to chlorides (version 1.0.0)*. University of Toronto, Toronto.

Betongföreningen (2007), *Guidelines for Durability Design of Concrete Structures* (in Swedish). Swedish Association of Concrete, Report No. 12.

Bigas, J.-P. (1994), *Diffusion of Chloride Ions through Mortars* (in French). Doctoral thesis, INSA-Genie Civil, LMDC, INSA de Toulouse, France.

Blunk, G., Gunkel, P. and Smolczyk, H.-G. (1986), On the distribution of chloride between the hardening cement paste and its pore solution. In: *Proceedings of the 8th International Congress on the Chemistry of Cement*, Brazil, Vol. V, pp. 85–89.

Buenfeld, N.R., Shurafa-Daoudi, M.-T. and McLoughlin, I.M. (1995), Chloride transport due to wick action in concrete. In: *Proceedings of the RILEM International Workshop on Chloride Penetration into Concrete*, 15–18 October 1995, St. Rémy-lès-Chevreuse (eds L.-O. Nilsson and J.P. Ollivier), pp. 315–324.

Byfors, K. (1990), *Chloride-initiated Reinforcement Corrosion: Chloride Binding*. CBI Report 1:90. Swedish Cement and Concrete Research Institute, Stockholm.

Castellote, M. and Andrade, C. (2005), *WP 2 Report: Pre-evaluation of Different Test Methods, ChlorTest: Resistance of Concrete to Chloride Ingress – From Laboratory Tests to In-field Performance*. EU Project (5th FP GROWTH), G6RD-CT-2002-00855, Deliverables 5–8.

Castellote, M. and Andrade, C. (2006), Round-robin test on methods for determining chloride transport parameters in concrete. *Materials and Structures*, 39(10):955–990.

Castellote, M., Andrade, C. and Alonso, C. (2001a), Relationship between Coulomb and migration coefficient of chloride ions for concrete in a steady-state chloride migration test. *Magazine of Concrete Research*, 53(1):13–24.

Castellote, M., Andrade, C. and Alonso, C. (2001b), Measurement of the steady and non-steady-state chloride diffusion coefficients in a migration test by means of monitoring the conductivity in the anolyte chamber comparison with natural diffusion tests. *Cement and Concrete Research*, 31(10):1411–1420.

Chatterji, S. (1994), Transportation of ions through cement based materials – Part 1: Fundamental equations and basic measurement techniques. *Cement and Concrete Research*, 24(5):907–912.

Collepardi, M. (1995), Quick method to determine free and bound chloride in concrete. In: *Proceedings of the RILEM International Workshop on Chloride Penetration into Concrete*, 15–18 October 1995, St. Rémy-lès-Chevreuse (eds L.-O. Nilsson and J.P. Ollivier), pp. 10–16.

Collepardi, M. and Biagini, S. (1989), Effect of water/cement ratio, pozzolanic addition and curing time on chloride penetration into concrete. In: *Proceedings of the ERMCO '89*, June 1989, Stavanger, pp. 606–615.

Collepardi, M., Marcialis, A. and Turriziani, R. (1970), The kinetics of penetration of chloride ions into the concrete. *Il Cemento*, 4:157–164.

Collepardi, M., Marcialis, A. and Turriziani, R. (1972), Penetration of chloride ions into cement pastes and concrete. *Journal of the American Ceramic Society*, 55(10):534–535.

Crank, J.C. (1975), *The Mathematics of Diffusion*, 2nd edn. Oxford University Press, Oxford.

Delagrave, A., Marchand, J., Pigeon, M., Ollivier, J.P. and Samson, E. (1996), Diffusion of chloride ions in high performance mortar. In: *High-Strength Concrete: An International Perspective* (ed. J.A. Bickley). *ACI SP-167*, pp. 175–191.

Denarié, E., Conciatori, D. and Brühwiler, E. (2003), Effect of micro climate on chloride penetration into reinforced concrete. In: *6th CANMET/ACI International Conference on Durability of Concrete*, Thessalonika.

DIN EN 206-1 (2001), *Concrete – Part 1: Specification, Performance, Production and Conformity*. Deutsches Institut für Normung, Berlin.

Dhir, R.K., Jones, M.R., Ahmed, H.E.H. and Seneviratne, A.M.G. (1990), Rapid estimation of chloride diffusion coefficient in concrete. *Magazine of Concrete Research*, 42(152):177–185.

Edvardsen, C. (1995), Chloride penetration into cracked concrete. In: *Proceedings of the RILEM International Workshop on Chloride Penetration into Concrete*, 15–18 October 1995, St. Rémy-lès-Chevreuse (eds L.-O. Nilsson and J.P. Ollivier), pp. 243–249.

EN 13396 (2004), *Products and Systems for the Protection and Repair of Concrete Structures – Test Methods – Measurement of Chloride Ion Ingress*. European Committee for Standardisation, Brussels.

EN/TS 12390-11 (2010), *Testing Hardened Concrete – Part 11: Determination of the Chloride Resistance of Concrete, Unidirectional Diffusion*. European Committee for Standardisation, Brussels.

Engelund, S., Edvardsen, C. and Mohr, L. (2000), *General Guidelines for Durability Design and Redesign, DuraCrete: Probabilistic Performance based Durability Design of Concrete Structures*. EU Project (Brite EuRam III), BE95-1347, Report R 15.

Feldman, R.F., Chan, G.W., Brousseau, R.J. and Tumidajski, P.J. (1994), Investigation of the rapid chloride permeability test. *ACI Materials Journal*, 91(3): 246–255.

fib (2006), *Model Code for Service-life Design*. fib Bulletin 34, 1st edn. Federation International du Beton, Lausanne.

Francy, O. (1998), *Modélisation de la Pénétration des Ions Chlorures dans les Mortier Partiellement Saturés en Eau*. Doctoral thesis, LMDC, Université Paul Sabatier, Toulouse.

Frederiksen, J.M. (1992), APM 302: Danish test method for the chloride ingress into concrete (in Danish). *Dansk Beton*, 2:22–27.

Frederiksen, J.M. (2000), Chloride threshold values for service life design. In:

Proceedings of the 2nd International RILEM Workshop on Testing and Model-ling the Chloride Ingress into Concrete, September 2000, Paris (eds C. Andrade and J. Kropp), RILEM PRO 19, pp. 397–414.

Frederiksen, J. M., Nilsson, L.-O., Sandberg, P., Poulsen, E., Tang L. and Andersen, A. (1997a), *A System for Estimation of Chloride Ingress into Concrete – Theoretical Background*. HETEK Report No. 83, Danish Road Directorate.

Frederiksen, J.M., Sørensen, H.E., Andersen, A. and Klinghoffer, O. (1997b), *The Effect of the w/c Ratio on Chloride Transport into Concrete: Immersion, Migration and Resistivity Tests*. HETEK Report No. 54, Danish Road Directorate.

Frederiksen, J.M., Mejlbro, L. and Nilsson, L.-O. (2008), *Fick's 2nd law: Complete Solutions for Chloride Ingress into Concrete – With Focus on Time Dependent Diffusivity and Boundary Condition*. Report TVBM-3146, Division of Building Materials, Lund University, Lund, Sweden.

Gautefall, O., Havdahl, J. and Justnes, H. (1995), *Determination of Chloride Diffusion Coefficient in Concrete from Migration Experiments: An Evaluation of the Method NT Build 355*. SINTEF Report, STF70 A94108, Trondheim, Norway.

Gehlen, C. (2000), *Probabilistische Lebensdauerbemessung von Stahlbetonbauwerken: Zuverlässigkeitsbetrachtungen zur wirksamen Vermeidung von Bewehrungskorrosion*. Heft 510 DafStb, Germany.

Gehlen, C. and Ludwig, H.M. (1999), *Compliance Testing for Probabilistic Design Purposes, DuraCrete: Probabilistic Performance Based Durability Design of Concrete Structures*. EU Project (Brite EuRam III), BE95-1347, Report R 8.

Gjørv, O.E. (2009), *Durability Design of Concrete Structures in Severe Environments*. Taylor & Francis, Oxford.

Gulikers, J. (2004), Critical evaluation of service life models applied on an existing marine concrete structure. In: *NORECON Seminar 2004: Repair and Maintenance of Concrete Structures*, 19–20 April 2004, Copenhagen.

Hansson, C.M., Strunge, H., Markussen, J.B. and Frølund, T. (1985), The effect of cement type on the diffusion of chloride. *Nordic Concrete Research*, 4:70–80.

Hassanzadeh, M. (2007), Internal testing report. Vattenfall Research and Development AB.

Holt, E. (2004), *Deliverable Report 8: Identification of Damage Mechanisms and Creation of Theoretical Models for Failure Phenomena, ConLife: Life-Time Prediction of High-Performance Concrete with Respect to Durability*. EU Project (5th FP GROWTH), G5RD-CT-2000-00346.

Hosokawa, Y., Yamada, K., Johannesson, B. and Nilsson, L-O. (2008), A development of a multi-species mass transport model considering thermodynamic phase equilibrium. ConMod'08, 2nd International Symposium on Concrete Modelling, Delft 26–28 May 2008.

ISO 5725-2 (1994), *Accuracy (Trueness and Precision) of Measurement Methods and Results – Part 2: Basic Method for the Determination of Repeatability and Reproducibility of a Standard Measurement Method*. International Organisation for Standardisation, Geneva.

ISO Guide 98-3 (2008), *Uncertainty of Measurement – Part 3: Guide to the Expression of Uncertainty in Measurement (GUM)*. International Organisation for Standardisation, Geneva

Johannesson, B. (2000), *Transport and Sorption Phenomena in Concrete and Other Porous Media*. PhD thesis, Report TVBM-1019, Division of Building Materials, Lund Institute of Technology, Lund.

Khitab, A., Lorente, S. and Ollivier, J.P. (2004), Chloride diffusion through saturated concrete: numerical and experimental results. In : *Advances in Concrete through Science and Engineering, RILEM Symposium*, 22–24 March, Evanston, IL.

Larsen, C.K. (1998) *Chloride Binding in Concrete. Effect of Surrounding Environment and Concrete Composition*. PhD thesis, Department of Structural Technology, NTNU, Trondheim.

Lindvall, A. (2001), *Environmental Actions and Response: Reinforced Concrete Structures Exposed in Road and Marine Environments*. Licentiate thesis, Publication No. P-01:3, Department of Building Materials, Chalmers University of Technology, Gothenburg.

Lindvall, A. (2003), *Environmental Actions on Concrete Exposed in Marine and Road Environments and its Response: Consequences for the Initiation of Chloride Induced Reinforcement Corrosion*. PhD thesis, Publication P-03:2, Department of Building Technology – Building Materials, Chalmers University of Technology, Gothenburg.

Maage, M., Poulsen, E., Vennesland, Ø. and Carlsen, J.E. (1995), *Service Life Model for Concrete Structures Exposed to Marine Environment: Initiation Period*. LIGHTCON Report No. 2.4, STF70 A94082 SINTEF, Trondheim, Norway.

Madsen, H.O., Krenk S. and Lind, N.C. (1986), *Method of Structural Safety*. Prentice Hall, Englewood Cliffs, NJ.

Mangat, P.S. and Molloy, B.T. (1994), Predicting of long term chloride concentration in concrete. *Materials and Structures*, 27(6):338–346.

Marchand, J., Samson, E., Maltais, Y., Lee, R.J. and Sahu, S. (2002), Predicting the performance of concrete structures exposed to chemically aggressive environment. Field validation. *Materials and Structures*, 35(254):623–631.

Masi, M., Colella, D., Radaelli, G. and Bertolini, L. (1997), Simulation of chloride penetration in cement-based materials. *Cement and Concrete Research*, 27(10):1591–1601.

McCarter, W.J., Emerson, M. and Ezirim, H. (1995), Properties of concrete in the cover zone: developments in monitoring techniques. *Magazine of Concrete Research*, 47(172):243–251.

McCarter, W.J., Ezirim, H. and Emerson, M. (1996), Properties of concrete in the cover zone: water penetration, sorptivity and ionic ingress. *Magazine of Concrete Research*, 48(176):149–156.

McGrath, P.F. and Hooton, R.D. (1996), Influence of voltage on chloride diffusion coefficients from chloride migration tests. *Cement and Concrete Research*, 26(8):1239–1244.

McLoughlin, I.M. (1997), *Modelling of Chloride and Moisture Transport In Concrete*. PhD thesis, Department of Civil Engineering, Imperial College, London.

Meijers, S.J.H. (2003), *Computational Modelling of Chloride Ingress in Concrete*. PhD thesis, Delft University, Delft.

Mejlbro, L. (1996), *The Complete Solution of Fick's Second Law of Diffusion with Time-dependent Diffusion Coefficient and Surface Concentration. Durability of Marine Concrete Structures*. Cementa AB, Stockholm.

Millard, S.G., Harrison, J.A. and Edwards, A.J. (1990), Measurement of the electrical resistivity of reinforced concrete structures for the assessment of corrosion risk. *British Journal of Non-Destructive Testing*, 31(11):617–621.

Nanukuttan, S.V. (2007), *Development of a New Test Protocol for Permit Ion Migration Test*. PhD thesis, Queen's University of Belfast, Belfast.

Nanukuttan, S.V., Basheer, P. A. M., McCarter, W. J. and Robinson, D. J. (2004), The long term effect of surface coatings on chloride penetration in concrete at different exposure conditions, in Proceedings of the 2004 symposium of Bridge Engineering Research in Ireland, Nov 2004, pp. 27–36.

Nanukuttan, S.V., Basheer, P.A.M. and Robinson, D.J. (2007), Determining the chloride diffusivity of concrete *in situ* using Permit ion migration test. In: *Proceedings of the Concrete Platform Conference*, April 2007, Belfast, pp. 217–228.

Nanukuttan, S.V., Basheer, L., McCarter, W.J., Robinson, D.J. and Basheer, P.A.M. (2008), Full-scale marine exposure tests on treated and untreated concretes: initial seven year results, *ACI Materials Journal*, 105(1):81–87.

Nanukuttan, S.V., Basheer, P.A.M. and Robinson, D.J. (2009), Development of a rapid *in-situ* ion migration test and comparison with the ASTM rapid chloride permeability test. In: *Proceedings of 7th International Symposium on Non-Destructive Testing in Civil Engineering, NDTCE'09*, 30 June to 3 July 2009, Nantes, pp. 523–528.

Nanukuttan, S., Basheer, P.A.M., Holmes, N., Tang, L. and McCarter, J. (2010), Use of performance specification and predictive model for concretes exposed to a marine environment. In: *Structural Faults and Repair Conference 2010*, 15–17 June, Edinburgh.

Nilsson, L.-O. (1992), *A Theoretical Study on the Effect of Non-linear Chloride Binding on Chloride Diffusion Measurements in Concrete*. Publication P-92:13, Division of Building Materials, Chalmers University of Technology, Gothenburg.

Nilsson, L.-O. (1993), Penetration of chlorides into concrete structures: An introduction and some definitions, in Chloride Penetration into Concrete Structures: Nordic Mini-seminar, ed. by L.-O. Nilsson, Division of Building Materials, Chalmers University of Technology, Gothenburg Publication P-93:1, pp. 7–17.

Nilsson, L.-O. (1997), A model for convection of chloride, chapter 7 in Frederiksen *et al.* (1997).

Nilsson, L.-O. (2000), A numerical model for combined diffusion and convection of chloride in non-saturated concrete. Proceedings of the 2nd International RILEM Workshop on Testing and Modelling the Chloride Ingress into Concrete, Paris, September 2000, RILEM PRO 19, ed. C. Andrade and J. Kropp, pp. 261–275.

Nilsson, L.-O. (2001), Prediction models for chloride ingress and corrosion initiation in concrete structures. *Nordic Mini Seminar & fib TG 5.5 Meeting*, 22–23 May 2001, Göteborg. Publication P-01:6, Department of Building Materials, Chalmers University of Technology, Gothenburg.

Nilsson, L.-O. (2005), WP 4 Report: Modelling of chloride ingress, ChlorTest: Resistance of concrete to chloride ingress – from laboratory tests to in-field performance, EU-Project (5th FP GROWTH) G6RD-CT-2002-00855, Deliverables 14-15.

Nilsson, L.-O., Massat, M. and Tang, L. (1994), The effect of non-linear chloride binding on the prediction of chloride penetration into concrete structures, in Durability of Concrete, ACI SP-145, ed. V.M. Malhotra, pp. 469–486.

Nilsson, L., Poulsen, E., Sandberg. P., Sorensen, H.E. and Klinghoffer, O. (1996), HETEK, Chloride penetration into concrete, State-of-the-art-report, Transport processes, corrosion initiation, test methods and prediction models, Report No. 53, Danish Road Directorate.

NT BUILD 208 (1996), *Concrete, Hardened: Chloride Content by Volhard Titration*. Nordtest, Espoo.

NT BUILD 355 (1997), *Concrete, Mortar and Cement Based Repair Materials: Chloride Diffusion Coefficient from Migration Cell Experiments*. Nordtest, Espoo.

NT BUILD 443 (1995), *Concrete, Hardened: Accelerated Chloride Penetration*. Nordtest, Espoo.

NT BUILD 492 (1999), *Concrete, Mortar and Cement Based Repair Materials: Chloride Migration Coefficient from Non-steady State Migration Experiments*. Nordtest, Espoo.

Ohama, Y., Demura, K. and Satoh, J. (1995), Behavior of chloride ions in unmodified and polymer-modified mortars, Proceedings of the international Conference on Concrete under Severe Conditions, August 1995, Sapporo, Japan, E&FN Spon, London, pp. 676–686.

Olesen, R. (1992), *PROBAN User's Manual*. Technical report, Det Norske Veritas, Høvik.

Otsuki, N., Nagataki, S. and Nakashita, K. (1992), Evaluation of $AgNO_3$ solution spray method for measurement of chloride penetration into hardened cementitious matrix materials. *ACI Materials Journal*, 89(6):587–592.

Page, C.L. and Vennesland, O. (1983), Pore solution composition and chloride binding capacity of silica fume cement pastes. *Materials and Structures*, 1:19–25.

Page, C.L., Short, N.R. and Tarras, A. El (1981), Diffusion of chloride ions in hardened cement pastes. *Cement and Concrete Research*, 11(3):395–406.

Pereira, C.J. and Hegedus, L.L. (1984), Diffusion and reaction of chloride ions in porous concrete. In: *Proceedings of the 8th International Symposium on Chemical Reaction Engineering*, Edinburgh, , Publication Series No. 87, pp. 427–438.

Petre-Lazar, I. (2000), *Evaluation of the Service Behaviour of Reinforced Concrete Structures Subject to Corrosion of Steel* (in French). PhD thesis, EDF, France.

Petre-Lazar, I., Heinfling, G., Marchand, J. and Gérard, B. (2000), Application of probabilistic methods to analysis of behavior of reinforced concrete structures affected by steel corrosion. In: *Proceedings of the 5th CANMET/ACI International Conference on Durability of Concrete*, 4–9 June 2000, Barcelona (ed. V.M. Malhotra). *ACI SP-192*, pp. 557–572.

Petre-Lazar, I., Abdou, L., Franco, C. and Sadri, I. (2003), THI: a physical model for estimating the coupled transport of heat, moisture and chloride ions in concrete. In: *2nd International RILEM Workshop on Life Prediction and Aging Management of Concrete Structures*, Paris, 5–6 May 2003.

Polder, R.B. (2001), Test methods for on site measurement of resistivity of concrete: a RILEM TC-154 technical recommendation. *Construction and Building Materials*, 15(2–3):125–131.

Poulsen, E. (1990), *The Chloride Diffusion Characteristics of Concrete*. Nordic Concrete Research Publication No. 9, pp. 124–133.

Poulsen, E. (1996), Estimation of chloride ingress into concrete and prediction of service lifetime with reference to marine RC structure. In: *Durability of Concrete in Saline Environment*, pp. 113–126. Cementa AB, Stockholm.

Ramachandran, V.S., Seeley, R.C. and Polomark, G.M. (1984), Free and combined chloride in hydrating cement and cement components. *Materials and Structures*, 17(100):285–289.

Richartz, W. (1969), The combining of chloride in the hardening of cement. *Zement-Kalk-Gips*, 22(10):447–456.

Romer, M. (2005), Recommendation of RILEM TC 189-NEC 'Non-destructive evaluation of the concrete cover': Comparative test – Part I – Comparative test of 'penetrability' methods. *Materials and Structures*, 38(284):895–906.

Roy, D.M., Kumar, A. and Rhodes, J.P. (1986), Diffusion of chloride and cesium ions in Portland cement pastes and mortars containing blast furnace slag and fly ash. In: *2nd International Conference on the Use of Fly Ash, Silica Fume, Slag and Natural Pozzolans in Concrete*, Madrid, ACI SP-91, pp. 1423–1444.

Salta, M.M., Melo, A.P., Gaecia, N., Pereira, E.V. and Ribeiro, A.B. (2006), Chloride transport in concrete from lab test to *in situ* performance. In: *Proceedings of International RILEM Workshop on Performance Based Evaluation and Indicators for Concrete Durability*, 19–21 March 2006, Madrid (eds V. Baroghel-Bouny, C. Andrade, R. Torrent and K. Scrivener), RILEM Pro 47, pp. 229–240.

Samson, E. and Marchand, J. (1999), Numerical solution of the extended Nernst–Planck model. *Journal of Colloid and Interface Science*, 215:1–8.

Sandberg, P. (1996), Systematic collection of field data for service life prediction of concrete structures. In: *Durability of Concrete in Saline Environment*. Cementa AB, Stockholm.

Sandberg, P. and Larsson, J. (1993), Chloride binding in cement pastes in equilibrium with synthetic pore solutions as a function of [Cl] and [OH]. In: *Chloride Penetration into Concrete Structures: Nordic Mini-seminar* (ed. L.-O. Nilsson), Publication P-93:1, pp. 98–107. Division of Building Materials, Chalmers University of Technology, Gothenburg.

SIA 262/1-B Chloride ions penetration coefficient.

Stanish, K.D., Hooton, R.D. and Thomas, M.D.A. (2000), *Testing the Chloride Penetration Resistance of Concrete: A Literature Review*. FHWA Contract DTFH61-97-R-00022, University of Toronto, Toronto.

Takewaka, K. and Matsumoto, S. (1988) Quality and cover thickness of concrete based on the estimation of chloride penetration in marine environments. In: *Proceedings of the 2nd International Conference on Concrete in a Marine Environment* (ed. V.M. Malhotra), ACI SP-109, pp. 381–400.

Tang, L. (1996a), Electrically accelerated methods for determining chloride diffusivity in concrete. *Magazine of Concrete Research*, 48(176):173–179.

Tang, L. (1996b), *Chloride Transport in Concrete: Measurement and Prediction*. Doctoral thesis, Publication P-96:6, Department of Building Materials, Chalmers University of Technology, Gothenburg.

Tang, L. (1997), *Chloride Penetration Profiles and Diffusivity in Concrete under Different Exposure Conditions*. Publication P-97:3, Department of Building Materials, Chalmers University of Technology, Gothenburg.

Tang, L. (2003a) *Chloride Ingress in Concrete Exposed to Marine Environment: Field Data up to 10 Years Exposure*. SP Report 2003:16. SP Swedish National Testing and Research Institute, Borås.

Tang, L. (2003b), *A Collection of Chloride and Moisture Profiles from the Träslövsläge Field Site: From 0.5 up to 10 Years Investigations*. Publication P-03:3, Department of Building Technology, Building Materials, Chalmers University of Technology, Gothenburg.

Tang, L. (2004), Modelling of chloride ingress in HPC. In: *Proceedings of International Conference on Durability of HPC and Final Workshop of ConLife*, 23–24 September 2004, Essen (eds M.J. Setzer and S. Palecki), Aedificatio, Essen, pp. 289–300.

Tang, L. (2005), *WP5 Report: Final Evaluation of Test Methods, ChlorTest: Resist-*

ance of Concrete to Chloride Ingress – From Laboratory Tests to In-field Performance. EU Project (5th FP GROWTH), G6RD-CT-2002-00855, Deliverables 16–19.

Tang, L. (2006), *In Situ Testing of Chloride Ingress in Concrete with Spacers* (in Swedish). Internal report, SP Swedish National Testing and Research Institute, Borås.

Tang, L. (2007), Service-life prediction based on the rapid migration test and the ClinConc model. In: *RILEM Proceedings PRO 047: Performance Based Evaluation and Indicators for Concrete Durability* (eds V. Baroghel-Bouny *et al.*), pp. 157–164. RILEM Publications, Bagneux.

Tang, L. (2008), Engineering expression of the ClinConc model for prediction of free and total chloride ingress in submerged marine concrete. *Cement and Concrete Research*, 38(8–9):1092–1097.

Tang, L. and Gulikers, J. (2007), On the mathematics of time-dependent apparent chloride diffusion coefficient in concrete. *Cement and Concrete Research*, 37(4):589–595.

Tang, L. and Hassanzadeh, M. (2009), Modelling of service life for a concrete structure exposed to seawater for 30 years. In: *Proceedings of International RILEM Conference on Concrete Durability and Service Life Planning*, 5–7 September 2009, Haifa. RILEM PRO 66, pp. 155–163.

Tang, L. and Nilsson, L.-O. (1991), Chloride binding capacity, penetration and pore structure of blended cement pastes with slag and fly ash. In: *Proceedings of the International Conference on Blended Cements in Construction*, September, Sheffield (ed. R.N. Swamy), pp. 377–388. Elsevier Applied Science, Oxford.

Tang, L. and Nilsson, L.-O. (1992), Rapid determination of chloride diffusivity of concrete by applying an electric field. *ACI Materials Journal*, 49(1):49–53.

Tang, L. and Nilsson, L.-O. (1993a), A rapid method for measuring chloride diffusivity by using an electrical field. In: *Chloride Penetration into Concrete Structures: Nordic Mini-seminar*, January 1993 (ed. L.-O. Nilsson), Publication P-93:1, pp. 26–35. Division of Building Materials, Chalmers University of Technology, Gothenburg.

Tang, L. and Nilsson, L.-O. (1993b), Chloride binding capacity and binding isotherms of OPC pastes and mortars. *Cement and Concrete Research*, 23(2):347–353.

Tang, L. and Nilsson, L.-O. (1994), A numerical method for prediction of chloride penetration into concrete structures. In: *The Modelling of Microstructure and its Potential for Studying Transport Properties and Durability, NATO/RILEM Workshop on the Modeling of Microstructure and its Potential for Studying Transport Properties and Durability*, 10–13 July 1994, St. Remy-lés-Chevreuse, (eds H. Jennings *et al.*), pp. 539–552. Kluwer, Dordrecht.

Tang, L. and Nilsson, L.-O. (1995), Chloride binding isotherms – an approach by applying modified BET theory. In: *Proceedings of the RILEM International Workshop on Chloride Penetration into Concrete*, 15–18 October 1995, St. Rémy-lès-Chevreuse (eds by L.-O. Nilsson and J.P. Ollivier), pp. 36–42.

Tang, L. and Nilsson, L.-O. (1996), Service life prediction for concrete structures under seawater by numerical approach. In: *Proceedings of the 7th International Conference on the Durability of Building Materials and Components*, 19–23 May 1996, Stockholm, pp. 97–106. E&FN Spon, London.

Tang, L. and Sørensen, H.E. (1998), *Evaluation of the Rapid Test Methods for*

Measuring the Chloride Diffusion Coefficients of Concrete. SP Report 1998:42, SP Swedish National Testing and Research Institute, Borås.

Tang, L. and Sørensen, H.E. (2001), Precision of the Nordic test methods for measuring the chloride diffusion/migration coefficients of concrete, Materials and Structures, 34 (242):479–485.

Tang, L. and Utgenannt, P. (2000), Characterization of chloride environment along a highway. In: *5th International Conference on Durability of Concrete*, Barcelona, 4–9 June 2000, Supplementary Papers, pp. 213–223.

Tang, L. and Utgenannt, P. (2007), *Chloride Ingress and Reinforcement Corrosion in Concrete under De-icing Highway Environment: A Study after 10 Years' Field Exposure.* SP Report 2007:76, SP Technical Research Institute of Sweden, Borås.

Theissing, E.M., Hest-Wardenier, P.V. and de Wind, G. (1978), The combining of sodium chloride and calcium chloride by a number of different hardened cement pastes. *Cement and Concrete Research*, 8(6):683–692.

Thomas, M.D.A. and Bentz, E.C. (2001), *Life 365: Computer Program for Predicting the Service Life and Life-cycle Costs of Reinforced Concrete Structures Exposed to Chlorides.* User Manual (version 1.0.0). Presented at the Nordic Mini Seminar & *fib* TG 5.5 Meeting, Göteborg, 22–23 May 2001.

Tritthart, J. (1989a), Chloride binding in cement: I. Investigations to determine the composition of pore water in hardened cement. *Cement and Concrete Research*, 19(4):586–594.

Tritthart, J. (1989b), Chloride binding in cement: II. The influence of the hydroxide concentration in the pore solution of hardened cement paste on chloride binding. *Cement and Concrete Research*, 19(5):683–691.

Truc, O. (2000), *Prediction of Chloride Penetration into Saturated Concrete. Multispecies Approach.* Doctoral thesis, Chalmers University of Technology and the National Institute of Applied Science, Gothenburg and Toulouse.

Truc, O., Ollivier, J.P. and Carcassès, M. (2000), A new way for determining the chloride diffusion coefficient in concrete from steady state migration test. *Cement and Concrete Research*, 30(2):217–226.

Tuutti, K. (1982), *Corrosion of Steel in Concrete.* Report Fo 4.82, Swedish Cement and Concrete Research Institute (CBI), Stockholm.

Uji, K., Matsuoka, Y. and Maruya, T. (1990), Formulation of an equation for surface chloride content due to permeation of chloride. In: *Proceedings of the Third International Symposium on Corrosion of Reinforcement in Concrete Construction.* Elsevier Applied Science, London.

Visser, J.H.M., Gaal, G.C.M. and de Rooij, M.R (2002), Time dependency of chloride diffusion coefficients in concrete. In: *3rd RILEM Workshop Testing and Modelling Chloride Ingress into Concrete*, 9–10 September 2002, Madrid.

Volkwein, A. (1991), *Investigations on the Penetration of Water and Chlorides in Concrete* (in German). Doctoral thesis, Munich University of Technology, Munich.

Whiting, D. (1981), *Rapid Measurement of the Chloride Permeability of Concrete.* FHWA/RD-81/11, Federal Highway Administration, Washington, DC.

Wirje, A. and Offrell, P. (1996), *Mapping of Environment: Chloride Penetration along National Road 40* (in Swedish). Report TVBM-7106, Lund Institute of Technology, Lund.

Zhang, T. and Gjørv, O.E. (1996), Diffusion behavior of chloride ions in concrete. *Cement and Concrete Research*, 26(6):907–917.

Index